HERDING CATS AND CODERS
SOFTWARE DEVELOPMENT FOR NON-TECHIES

BY GREG ROSS-MUNRO

Herding Cats and Coders Software Development for Non-Techies

Greg Ross-Munro

Copyright © 2018 by Greg Ross-Munro

ISBN-978-0-9986801-1-8

In memory of Glyn Davies, my grandfather.

Who taught me to draw a pig in Microsoft Paint. And then somehow felt that binary division was the next logical lesson for an eight-year-old.

CONTENTS

FOREWORD

by Marc Canter

As entrepreneur and software engineer Marc Andreessen has famously said, "All companies are now software companies." In other words, software is eating the world. There is no way around it—without a fundamental understanding of how software systems work, and how technology affects your company, you are doomed.

Many have tried to ignore this and have suffered for it. Remember that Target credit card breach back in 2013? It cost them over $2 billion and affected 70 million customers. Ever heard of Toys "R" Us? They thought they could "buck" the Internet. They believed they didn't need a website or online distribution! How'd that turn out for them? Whether it be for process control, advanced accounting and big data analysis, or automated marketing campaigns, technology is pervasive in the business world today. It all leads back to "software." But what does that word even mean?

The main purpose of this book is to educate nontechnical C-level execs in the mysteries and inner workings of software development. By reading this book, anybody can understand not only the basic decision nodes, methodologies, and processes that come along with software technology development, but also the three-letter acronyms, technical jargon, and sophisticated language used by IT guys and pros hired from outside software development or consulting firms.

If tech-speak tends to go over your head, the good news is, you're not alone. Most nontechnical people experience this phenomenon. The bad news is, if you don't get a clue, you'll be pouring good money after bad, and you'll then have to hire MORE technical people—this time paying them to fix all the mistakes done the first time. (I'd say this happens over half the time when nontechnical managers and execs hire technical staff. Companies like Accenture, McKinsey, and IBM thrive on this dysfunctionality!)

Whether for productivity, personal career choices, or creativity, programmers are the thoroughbreds of our era—the rock stars. The "hacker," the technical person whose fingers move on the keyboard and who actually does the work—that's who's important! And you'd BETTER be able to hire them, manage them, and keep them happy if you hope to build a successful and fully functioning software system.

I have spent almost forty years immersed in building software, both as an employee and entrepreneurial/CEO guy. These days, the Internet is largely taken for granted as online infrastructure, but it wasn't so easy "back in the day." I'm so old that "Save As…" and "Print" were fancy new features when I started. I used to sit on "software" panels with some guy from an accounting company and a gal representing "printer drivers." That's what the software industry used to be. Several downturns, crises, upturns, Y2Ks, booms, busts, and bubbles later, we're a multi-trillion dollar industry that there's no hiding from.

I founded a software company called MacroMind in 1984, which originally was going to build tools to enable artists, musicians, and designers to build video games. Along the way, this machine called the Macintosh appeared, and everything changed. We were in the right place at the right time, as it turned out that "multimedia" was also applicable to scientific visualizations, marketing animations, product simulations, and video games. At its height, 85 percent of the world's CD-ROMs were produced with our product: Director. Imagine that—the same product, able to be used by hundreds of thousands of nontechnical types, producing countless varieties of interactive content! This was all back "before" the Internet.

MacroMind would eventually become Macromedia in 1991, the world's leading multimedia tools company responsible for producing Director, Flash, and Dreamweaver software (later merged with Adobe Systems in 2005.) Since then, I've been involved in so many tech and software ventures that I can barely count them all. Within certain circles, I'm known as a serial entrepreneur, a tech evangelist, a UI and UX expert, and an outspoken advocate for users and user rights.

My focus has always been on making interactive web technology more accessible and more readily available to real people. I could argue that the purpose of software is to change the world, not make money. That those of us who figure out how to make money along the way—while doing what we love—are blessed. But that's just not how the world works. Today, you MUST understand software.

I first met Greg through a number of startup programs I was involved with, including the Gazelle Labs program, and I've come to know him as an expert in his field. As the founder and CEO of SourceToad, a B2B software company responsible for designing large-scale API management systems, Greg has spent years educating his clients on what he and his fellow developers do behind the scenes. In the following pages, he shares that information in a book that is devoted to demystifying the process of software development.

For geeks like me, this book is a mundane, simplistic summary of issues which are extremely important. But to NON-geeks—nontechnical staff and execs— EVERYTHING in this book is essential and must be understood. Period. No ifs, ands, or buts.

I especially liked Greg's section detailing a company's internal systems and the online tools required to produce software. I also really appreciated how Greg put everything "into context." You won't have to wonder what your developers are talking about when they say things like, "API," "UI," "UX," "wireframing," and "cross-platform applications." You will also learn the most effective strategies for communicating with developers and coders, including:

- What information/resources the development team needs from you in order to do their jobs well

- How to get involved in a development project without micromanaging

- How to select and implement a management methodology (thus circumventing a "cowboy coding" scenario)

- What makes a developer feel good about his work

- How to provide constructive feedback to your development team without pissing them off

This book gets the 80/20 rule of what's important for software development. It's a great intro for nontechnical types to educate them on all the issues involved in producing mobile apps, websites, and enterprise software. In fact, I'd go so far as to say that this book is not so much about herding cats as it is about getting a clue about why herding those cats is important!

I believe that in the battle of "man vs. machine," open information is power, and Greg's book provides just enough knowledge to be dangerous, so that anyone can venture out there and "dip their toe" into the world of software development. Actual experiences may vary, but I guarantee you that if you focus on:

- Hiring the best people
- Developing really good ideas that solve a real problem and deliver value and benefits to end users
- Getting your timing right, leveraging social media, testing the hell out of your software, validating your ideas and assumptions, and attending a few conferences or trade events

... Then you'll be successful!

A software development crash-course like Herding Cats and Coders ought to be required reading for any manager or director preparing to work with software developers or IT people in general. By getting yourself on the same page as the experts, you'll avoid problems and reduce stress, and your project will be better off for it—not to mention your budget.

This book offers an invaluable level of transparency for anyone who works with (or is planning to hire) programmers or programming teams. When you need a piece of software built for your business, the information shared here will help you ask the right questions and develop a strong, useful knowledge base. Never again will you have to scratch your head and wonder, "Just what on earth are those damn IT people doing, anyway. . . ?"

Marc Canter
Founder and CEO of Macromedia, Cantervision, Instigate, Broadband Mechanics, and "Godfather" of the multimedia industry
Walnut Creek, CA

ABOUT THIS BOOK

Picture this. You need to build a house. You can't build it yourself because you're not an expert, but you do have a basic understanding of the process. You know that you need to hire an architect or pick out your base floor plan, and you need to find a builder with a skilled crew, get the right permits, and pick out the land. Soon, the crew starts laying foundations. Studs, floors, plumbing, and wiring go up next, and the process continues.

You know all this because you've driven past thousands of construction sites in the past. You've seen the guys in hard hats at work. You also live in a house and visit other people's homes, so you have internalized most of the key features of "house-ness." You know what a house is, and you have an idea of how one is built.

But you can't drive past an app mid-build.

Despite the fact that most of us use software every day, the process of how it is created and built is almost completely hidden from view. There are no guys in hard hats for us to watch. So when businesses need to build an app or some other highly specialized tech service or product, they often don't know what to expect. They know they have to hire some experts, but they find themselves lacking a frame of reference in common with the people who will be doing the work.

I've been developing software since the age of eight. (It wasn't very good software, for sure, but it got better—I promise!) My first real job was building Internet software before the web existed, and for the last ten years, I've run a contract software engineering company, which now sports offices in the United States and Australia. We've built software for Procter & Gamble, Sony, Viking Cruise Lines, Jackson Hewitt, Honeywell, and even U.S. Special Forces.

Most of the clients who come through our doors are not programmers. (It would be weird if they were, right?) Their industries, companies, and startups rely heavily on technology, but they are not technologists. In fact, the vast majority of executives and managers involved in making decisions about software development budgets are non-experts.

Over the years, I found that this problem just kept coming up. Clients would enter our offices knowing what they needed, but they had a very limited understanding of what went on behind the scenes of tech development. The rapid changes to software in the last two decades have made it almost impossible to keep up unless it's your full-time job. I needed a resource I could share with my clients to help them better understand the process. I searched for something—a book, a training course, anything—that would give my new clients a jump on the education process required at the start of any new project. Obviously, I didn't find it. If I had, I would have bought a hundred copies and stacked the books in our storeroom. Instead, I had to write the thing myself. So, you're stuck with my version.

Software development has become an indispensable part of business. From the budding startup to the largest Fortune 500 enterprise, very few companies escaped the technological revolution of the last twenty years. Every organization in the world can benefit from a software solution of some kind, and due to the increasing complexity and specializations of companies, there just isn't something you can buy off-the-shelf to achieve those benefits. Most large companies now have in-house development teams or contract out custom work to specialized software studios. The problem is that software development is a highly specialized, highly technical field that is rich in jargon, obscure acronyms, and a myriad of technologies esoteric to the non-nerd.

Basically, I wrote this book because there was no easy way for non-engineers to get a broad overview of the concepts and processes of software engineering. This book is my attempt to solve that problem by closing the gap between what non-experts know about the development process and what a programmer actually does for a living. My theory on how to do this is simple:

1. Outline a few technical topics that come up in development conversations so you can avoid drowning in jargon.

2. Familiarize you with the process of software development from the early planning stages to the launch of a new system. We'll discuss how long projects take, how much they cost, and how to get the best value for your money.

3. Discuss a number of business-related topics to give you an overview of
    the software industry, the market, and the types of people with whom
    you'll work.

In each chapter of Herding Cats and Coders you will find:

- Concise summaries of technologies you should know

- Advice on how to make informed decisions on software investments

- Tips for dealing with programmers, digital departments, and outside
  software vendors

I've tried to collect everything you would need to know about creating and
launching a software product or service, short of programming the thing
yourself. This is not voodoo. There is no "engineering magic," regardless of what
any programmer or IT person tells you. Armed with this book, you will have an
executive overview of the software creation process, and you will be able to call
BS when you hear it from snarky engineers!

Who Is This Book For?

This is not a book for programmers. This book is for those who have to work
with programmers and programming teams to get software built for their
businesses.

Most people responsible for software projects are CEOs, marketing directors,
project managers, and entrepreneurs. Not being on the same page as your
development team leads to poor products, cost overruns, and project failures.
Herding Cats and Coders will put you on the same page as the techies. This is
the perfect book for a non-technical manager, whether working with an outside
developer, a development team, or an agency. And to avoid boring you to death
by turning this book into a dry, technical treatise, I've kept things lighthearted—
and irreverent at times.

If you are a non-technical user or manager who either works with developers or
plans on engaging with software engineers to build something, Herding Cats
and Coders will help you succeed.

How to Use This Book

This book was written to give you a head start on building any type of software. It serves as your primer on the three key elements of understanding the entire software production life cycle.

- The raw building blocks that make up software

- The processes involved in the production of software

- How to launch and market software products and services.

I'll focus specifically on working with teams of programmers, project managers, and designers to build an application for the web or a smartphone, but we'll touch on all types of software, and the methods discussed are universal.

After reading this book, you should be able to walk into a programmer's office, or through a development firm's door, and have a solid understanding of the technologies, processes, and terms that will be discussed. You will know how to choose a strong development team and how to hire skilled programmers. You will also have a much better idea when someone is trying to pull the wool over your eyes.

Herding Cats and Coders is broken up into three main parts:

Part 1: Getting Started

In this section, we'll do a broad overview of what an application is, how coding works, and what the general anatomy of an application looks like.

Part 2: Start Coding

In this section, we'll explore all aspects of building software. We'll start with what the possibilities are, then move on to how to decide what to build. After that, we will look into how to find developers to build your application (including how long it will take to build and how much it will cost). Finally, we'll dive deeper into the steps involved in creating software, from the planning process through to testing.

Part 3: Launching and Beyond

In the final section, we'll discuss what happens after your application has been completed. Just because the coding is finished does not mean the journey is over! We'll examine how applications are distributed, marketed, and monetized.

This is not a book about how to code, though there are a few code examples throughout the book. These examples are an attempt to be as illustrative as possible, while acknowledging that you're probably not going to start programming as soon as you put the book down (or even try following along with the examples). I designed the code in this book with your ability to read and understand in mind. If you manage development projects or hire developers, it would be a good idea to get into programming, at least a little. You can probably pick up the basics in a weekend with an online tutorial. Codecademy (https://www.codecademy.com/) has a few great free courses that will give you a feel for how code works. I promise you, it's easier than it looks!

Imagine running a team of accountants doing financial audits if you weren't an accountant yourself. It would be exceedingly difficult. Your direct reports probably wouldn't respect your opinion or leadership. But just five or six hours of your life spent learning the basics of software development will be enough to stand you in good stead.

And who knows? You might actually like it!

PART I
GETTING STARTED

CHAPTER 1
DECIDING WHAT TO BUILD

"The beginning is the most important part of the work." —Plato

Seeing as you are on your way to being an expert in the technologies used in software development, you're probably looking forward to working with your developers to build something awesome. But wait a minute! There is a whole world of caveats and stumbling blocks unique to software development (although they are very similar to every other industry out there). Step one is simply to determine what you're building and why.

1. Software is typically built for three main reasons:
    Software as a new business, product, or new business line: Here lies your typical tech startup out to change the world, or an existing business trying to launch a new game, product, or service. These are high-risk, high-reward buildouts.

2. Software as marketing: This is everything from a shiny, new website for your company to an app to help marketing efforts. These are low-risk projects, built to increase brand awareness, brand loyalty, public perception, or customer acquisition.

3. Process improvement software: These are internal projects run by existing organizations trying to enable their employees or customers to better work with the company's preexisting systems. This might be automating a process or building a mobile version of an older system.

In this section, I'll go over some basics of doing market research around a software idea. After that, we'll have a quick discussion on the dreaded topic of needs analysis. Afterward, maybe a break for coffee, or tea . . . or whatever.

To Build or Not To Build

"Once you make a decision, the universe conspires to make it happen." —*Ralph Waldo Emerson*

So, you've got a great idea or a huge problem that needs to be solved. Perhaps it is the inefficient system you work with every day. Any way you look at it, you know that if you built a software program, you could fix the issue! The problem is that designing custom software is a long, complicated, and expensive process. On top of that, it's never really complete. There will always be bugs, changes, maintenance, and a million little things that reading this book will prepare you to handle. My point is, you should only get into a custom development project once you know two things:

1. Is there anything out there that already does what I want to do?

2. Can I really build a solution that is at least three times as good as the nearest competitor or substitute?

Answer these two fundamental questions to make sure that, (1) you aren't reinventing the wheel, and (2) if you are redesigning the wheel, yours will be so superior that people will make the effort to switch.

Is there anything out there that already does what I want to do?

There are millions of pieces of software out there. The app stores grow more and more crowded every day. Even if you're searching for "hamster weight tracking software," the web will present you with at least twenty different options. It's a very competitive market. There are two ways to look at this question:

1. I am building something custom for my company to run more efficiently.

2. I am building something to sell to businesses or the public.

In both cases, you need to do a lot of research about your idea and the potential competitors or replacements. The first thing you'll have to do is get over the concept that your idea is unique. It's not. Well, almost certainly not. Your approach might be a unique take on an idea, but it's not novel.

If you're building the best hamster weight loss application ever, it would likely be the only one. There are thousands of weight loss apps out there, and, believe it or not, almost just as many featuring hamsters in some way. You owe it to yourself to look long and hard at all of them. Do hamsters need their own weight loss application? Could one designed for cats work equally well? Is there something really good already available? And would the developer work with you on adding a vertical?

Most of the applications I see come through my door are business process related. For example, a company may have a complicated delivery system that requires barcoding, approval, receipt of a packing number, sorting, and finally, approval again for each incoming package. Pretty boring. Also pretty important. Maybe I should have stuck with hamsters.

When a company like that comes to me, they almost always want us to build a custom system that allows their packers to use an app to do the scanning and checking in; some web application to let the managers know how everyone is doing in real time. Regardless of how unique the company's business processes are, some of them could probably get away with an off-the-shelf solution. Maybe what is currently available would only get them 70 percent of the way there, but a commercial application would probably come with support, a marketplace of consultants, and training.

You might be thinking that maybe you, too, could get away with a solution "out of the box," but the truth is that this cliché is a falsehood. In these situations, you replace custom development with custom integration. Huge software companies that sell "enterprise software" packages are very powerful, but they will leave you with two basic choices:

1. Spend large sums of cash on consultants, support contracts, and integrations specialists (a fancy term for developers).

2. Change your business processes to fit the software (rather than the other way around).

With a from-scratch custom build, you have a system that matches your process exactly and improves that process because of the thinking that has gone into it. The main question then becomes, "Are we really equipped and funded to support a fully customized build?" Whether it is a question of build or buy, the answer is fairly similar for the budget and ownership parts of the equation. The "are we equipped" part of the question is the most important.

Can I really build something that is at least three times better than the nearest competitor or substitute?

Why does it need to be *three times* better? Because customers have inertia.

People are lazy. They don't want to switch to your new social network. Facebook is good enough (and all my friends are already there). Smartphones had been around for ten years before Apple released the iPhone. Its release made people switch carriers, jumping to AT&T from their previous provider, something never before seen in the market.

This is a more complicated question than simply finding out if something already exists. We're now wandering into the realm of entrepreneurship, product design, and domain expertise. In fact, I'm going to take a cop-out answer right now and say that you probably can't know for sure if what you build will be three times better. This question is too complex. It requires instinct rather than spreadsheets. Market research helps. Talking to potential customers helps. But it's something you must feel in your bones.

Even if you can't answer this question for certain, there are a few basics with which you should become familiar. These are questions you should ask yourself about the application you're going to build. The first focuses on new businesses and startup entrepreneurs. If you are an existing business, skip to question two.

1. Am I going to get rich doing this?

The vast majority of my clients are businesses. They are looking to reduce inefficiencies, improve customer satisfaction, or delight their users in some way. Certainly, these businesses consider their bottom line, but money is not really a huge issue in terms of budget. For the 20 percent or so of our clients who are startups, almost all of them are non-techies, which is why they're coming to us for help. But what are their odds of making it big?

Every day, we read about tech startups raising millions of dollars in San Francisco to build companies we often don't understand. If they're making it big, why can't I? If you read TechCrunch (www.techcrunch.com), you'd think that billion-dollar ideas were a dime a dozen. Every day, they post articles about the next big thing, the next giant venture capital deal, and the next insane valuation of a twenty-something's startup.

Think of sites like TechCrunch as news outlets dedicated to covering shark attacks. Shark attacks, as rare as they are, are almost as frequent as multi-million dollar raises by institutional investors. According to the National Venture Capital Association (yes, there is an association for everything), 2014 was the most active year in VC spending in over a decade, with 1,799 tech-related deals. Sure, there were only 72 shark attacks in 2014, but we're still talking tiny numbers here. The vast majority of those were in San Francisco and the surrounding areas—the tech deals, not the shark attacks.

Judging by the data, you're probably not going to land a large, institutional investment unless you're under thirty, live in the Bay Area, and went to Stanford. That doesn't mean you can't do well in tech, though. A friend of mine made millions on a software company in Salt Lake City, of all places, and bootstrapped the entire business. So, if you're not a hoodie-wearing Stanford alumnus, where do you go from here? What questions should you ask yourself before you dive in?

2. Am I an expert in the space?

It continues to amaze me how few people ask themselves this question before getting started. If you go into business or you want to develop a product for a market, you have to know the business and market. This should be obvious, but I have found that, for whatever reason, it's not.

I am frequently invited to judge startup competitions, and time after time, I see pitches where the CEO had a good idea but clearly doesn't know anything about the industry. Tech, in general, suffers from this problem. In the media, we see companies breaking into new markets all the time, leading us to believe it's just a matter of building the right software and getting some marketing dollars. What we don't see is the long list of advisors, experts, and business connections that those companies have who have helped along the way.

If you're a sole proprietor who has run an ice cream shop for the past ten years, then decide to pull the trigger on your brilliant idea for a women's online clothing business, you're probably doing it wrong. Build an application that deals with ice cream!

The hidden problems with any business, market, or customer base are subtlety and authenticity. If you're not a domain expert in what you're doing, you may have a brilliant idea, but you don't know what you don't know. And those are the gaps that come back to bite you in the ass.

Would you buy a dress from a former ice cream salesman? Or would you rather buy from someone who has been in the industry for fifteen years? Even if it's only on the distribution side, you have a leg up on the ice cream guy. Do what you know. Ideas are a dime a dozen. If you don't know the space, come up with a different idea.

3. Am I in this for the long run?

The number one factor I've seen in the success of software projects and businesses is a commitment to the long-term success of the endeavor. Two years seems like the magic number where a startup begins to work or a new business line starts to take off. Two years is a really long time in business, and it's an even longer time if you're self-funding. That kind of commitment isn't for everyone. However, if you're serious, and someone telling you it takes two years to make a real go of it doesn't scare you, you're good to go!

If you've honestly asked yourself these questions and you're not terrified, it's a good sign that you are the right person or business to create something three times better than the competition.

Your Target Audience

"The real voyage of discovery consists not in seeking new lands but seeing with new eyes." —Marcel Proust

Any product or service starts with a person in mind. The more specific the better. Ask yourself very seriously, "Will anyone care about this product?"

The first step is to build a model of your perfect target user. When we do this, we always give the model target user a name. I know that sounds a little strange, but it really helps to personalize what you're doing. For internal projects, use the name of an actual employee who will use the application. That way, you'll never lose sight of the human factor. A typical user profile we use looks something like this:

User Profile for Medical Tracking App

José is a sixty-five-year-old man from Jacksonville, Florida. He works as a principal at a local middle school, where he loves his job and deals well with the insanity and drama that middle-schoolers impose on him every day. He has two children: a son in college in Miami and a daughter who lives close to home, working as a financial analyst.

José has recently been diagnosed with early-stage prostate cancer. At the recommendation of his physician, he downloaded the Medical Tracker app from the app store on his Android Samsung Galaxy S5. With the help of his daughter, Alexa, he will use the application to remind him when to take his medication, as well as to record any adverse effects he feels throughout the day. José is not really accustomed to downloading apps and doesn't use his phone for much other than phone calls and reading email. He never uses the phone to respond to email, but rather switches to his work PC, which runs Windows 7. He uses the default Internet Explorer 11 browser for web browsing and Microsoft Outlook for emails. He is comfortable in Microsoft Excel and Word, but that's about it.

He watches a lot of cooking shows on TV, as well as fanatically following the Jacksonville Jaguars in the NFL. He does not have a Netflix account, although his kids can't understand why.

We now feel like we understand the type of person who is going to use our application. It humanizes the software for developers and testers alike.

It's tough for a twenty-eight-year-old programmer to understand the life of a sixty-five-year-old middle school principal, so this is the first step to helping the designers and developers get into the right frame of mind. This affects everything—from the font size of the text to the contrast of the buttons. If you're designing an application for middle-school students (rather than their principal) you might not have any prompts or explanations as to how the app works. Older users may need more prompts and tutorials, especially if the software relates to something important like medical issues or finances.

Market Research

Once you're done setting up your perfect user, it's time to do a little target market research. Conduct competitive research by searching Google for the types of keywords you expect your audience will use to find your app. Whatever comes up is worth a look. The Internet is a very big place, so you are likely to find something that has most of what you're seeking to do. Look at screenshots, product demos, and YouTube videos of competing products, and feel free to steal their best ideas!

After you have a good idea of what's out there, start working out how your target audience is going to find your application. Here, you'll want to see what people are searching for online—it's the cheapest and easiest way to check your audience size. My two favorite systems for doing this are App Annie (https://www.appannie.com/) and Google Trends (https://www.google.com/trends/). Using these systems, you can enter your keywords to see how many people are searching for certain phrases and even related keywords. This helps you get a feeling for the size of your market and gives you some insight into any related fields you might want to consider for marketing purposes. You might even discover features you hadn't considered.

Real People and Surveys

Last but not least, you should try out your ideas on living, breathing people. You might want to set up some focus groups with friends and family or grab some of your employees who will interact with the end product. Stick them in a room with some of your early mockups and get feedback on what they think.

Remember that these folks aren't going to be experts. They are going to give you feedback that might be too broad. Common to this approach is the, "It would be cool if it could also do X" feedback. The person might not understand that you don't have all the time and money in the world to make the greatest Swiss Army Knife of an app out there. The best way to deal with this kind of feedback is to say, "Thank you! That's a great idea. I'll add that as a potential future roadmap item."

Finally, all feedback is good feedback. Don't take anything personally. The only response you'll ever need is, "Thank you, that's a great point." Even if you don't think it's a good point, you will want to drag these people back in to do some free QA (quality assurance) for you, so it's best to keep them happy. Bear in mind that it probably is good feedback, even if you disagree. If someone mentions something that is out of scope, too broad, or downright wrong, it can mean you haven't explained the purpose well enough, or that the point isn't obvious enough. Smile, say thank you, and make a note to address it later.

Another great tool is to set up a simple survey and get people you know on Twitter, LinkedIn, and Facebook to take it. Google forms (http://www.google.com/forms/) and Survey Monkey (https://www.surveymonkey.com/) are excellent resources for doing this.

Stakeholders

"In the long history of humankind (and animal kind, too) those who learned to collaborate and improvise most effectively have prevailed." —Charles Darwin

There are people involved in any project outside of your target audience. These are your stakeholders, the people who either pay for the development or become affected by it in some way. In larger companies, the most important stakeholders might be the CEO, the head of your department, or the board of directors. If you're an entrepreneur, your stakeholders are most likely your investors. These are investors of time, not just money. If you have a partner or spouse, they need to be on your side. There is nothing in the world more difficult than fighting for a business's survival out in the market, then fighting for it again when you get home. When it comes to financial investors, they will need to buy into the entire vision while understanding that the product side is equally important.

The old adage that investors should bring more than money is even truer for tech investors. Hopefully, you have an investor who knows the industry you're trying to break into, and not just some rich person keen to gamble.

Key Customers

Like your target audience, your most important stakeholders are your key customers. These are the people you know will use the application. Getting feedback from them early is pivotal to the success of any project. Many companies and entrepreneurs are hesitant to show their early-stage ideas to actual customers. This is due to fears that they won't like it or, worse yet, that they might steal the ideas.

Keep in mind that if a potential customer were really going to steal the idea, they would do it anyway. But it is much more likely that they have no interest in becoming a tech company. If you are so concerned that a customer is going to take something and do it on their own, you might be building something that doesn't bring the most important piece of value to the system: you. Your expertise, resources, or true grit are what will make or break a new development project. If that isn't well understood by the people you talk to early on, then they're most likely the wrong "key customer." Or you need to rethink your application entirely.

The Development Team

Your development team will be a group of stakeholders as well. Programmers are just like regular people (but don't tell them I said that), and everyone likes engagement in what they're doing. Getting your developers to not only understand your idea but to buy into it will go a long way toward getting the best out of them.

Contract developers have most likely worked on a wide variety of projects, many of which they saw little value in or didn't fully understand the purpose of. When a programming team is able to see the business need, grasp the utility, and understand your target audience, they are going to believe in what they're doing. Nothing is more powerful than when the people doing the actual work truly buy into that work.

A quick note about labels. . . . Most people in the non-tech world lump everyone who works with computers into a bucket called "IT." **Information Technology** is a broad term and was a fairly accurate way to describe computer nerds for a long time. Today, however, the lines are a little more clearly drawn because of the explosion in complexity in the number of systems out there. Networking, security, mobile development, web development, hosting, backup systems, payment solutions, and a myriad of other disciplines are so specialized that almost no one is an expert in more than one or two.

The clearest line in the sand is between the software developers and the hardware and network people. When you hear the term "IT" these days, it's usually referring to the hardware and network crowd. Software developers, or simply "tech," is the other faction. These two groups often do not get along. Be warned.

Your team—IT and otherwise—will be an invaluable source of honest feedback. Developers who have worked on numerous projects have seen all sorts of ideas play out in practice. A good team advises you on what worked in the past and the pitfalls to avoid. Take them out for a drink or get them to come on-site with you, and you'll take a big step toward getting your new team of nerds on your side. In the next section, we'll answer a question that has daunted non-nerds for years. . . . What is software, anyway?

CHAPTER 2
SOFTWARE AND SOFTWARE DEVELOPERS

What is Software?

Software is something almost everyone interacts with every day, but most people don't know much about how it all works behind the scenes.

Basically, software is the non-physical stuff that makes a computer work, or the intangibles that run on a computer. That means everything from the operating system to the word processor on which I'm writing this. Software also includes all the little files that help configure other pieces of software, like libraries and drivers.

At their core, computers are pretty stupid things, an assertion I'm sure you've expressed on more than one occasion. Computers are just really big calculators that handle lots of simple calculations to produce much more complex outputs. This includes everything from the ability to store files all the way to playing the latest video games. How do they do this?

Computers work on a simple system of mathematics called binary code. Binary is a way of representing numbers by using only 1s and 0s. Not to get too technical, but binary code represents all numbers, decimals, and fractions like this:

Binary Number	Decimal Number
0000	0
0001	1
0010	2
0011	3
0100	4

Nerd Out

The right-hand digit of a binary number represents the 1s column, the same as in our base 10 counting system. The second digit represents 2s, and the third represents 4s. They continue in this base-2 fashion: 1, 2, 4, 8, 16, 32, 64, 128, 256, 512, 1024, and so on. Those numbers probably look familiar to you if you've ever bought memory cards or looked at your computer monitor's display resolutions. That's because they are the incrementally larger sizes required to store information.

Computers contain something similar to a lookup table that enables the computer to represent all sorts of other information. This works a little like a decoder book, where you have a table of codes and a key to tell you what they mean. Think Morse code, but with a lot more options. For example, a computer represents the English alphabet like this:

Binary Number	English Alphabet
100 0001	A
100 0010	B
100 0011	C
100 0100	D

In this same way, we can represent specific points on a screen, or the exact tone that comes out of your speakers when you're listening to your iPod, or pretty much anything else that goes on in the daily life of a computer.

What we've described here is the "alphabet" of a computer; the most basic building blocks. Numbers aren't useful on their own unless you can add, subtract, multiply, and divide. Letters aren't valuable unless you can combine them into words. In order to make these basic building blocks effective, computers need a number of tools to manipulate these simplest forms. We call these tools "instructions."

On every computer in the world, there are a number of "processor" chips. We refer to the most powerful chip in the computer as the "central processing unit," or "CPU." Modern computers often have multiple processors strewn around to work on specific tasks. These are small, square, regularly black little things that plug into the main circuit board that connects to every other part of the computer. Some handle the graphics systems, the sound, or any other number of specialized tasks.

The CPU is akin to the brain of the computer. Modern processors are a marvel of modern engineering, but at the end of the day, they are pretty straightforward devices. Encoded in each processor is a set of instructions.

These are recipes the computer follows over and over again. These recipes could look something like this:

1. Check to see if anyone pressed a key on the keyboard.

2. If they did, make a note of what they pressed.

3. Send that key press to the screen.

4. Display that key to the user.

Or something like this:

1. Do I need to add two numbers? Yes!

2. Ok, please give me the two numbers. :)

3. From the keyboard: 2 and 5

4. Thanks! Here is the answer: 7

Computers don't normally have the cutesy nature of asking politely and responding with smiley faces and exclamation marks, but you get the idea. Processors these days have millions of these recipes, and most of them are much more complex than the two previous examples.

We group all of these recipes into what chip designers call "instruction sets." Instruction sets run continuously from beginning to end, while the computer is on, in the "instruction cycle." In the early days, computers were able to run one cycle at a time, but newfangled, fancy machines run multiple cycles at the same time. The more cycles the computer can run at the same time, and the quicker it can run them, the faster your computer runs. You've probably seen this speed printed on the box of every new computer you've ever bought. When you buy a computer that runs 2.5GHz, it means (very roughly) that your computer is doing 2.5 billion instruction cycles per second . . . which is pretty quick.

Building software is the process of stringing together recipes to create a larger, more useful whole. A single line of computer code may reference hundreds of instructions at the CPU level and only do something as simple as print some text to a screen. Programmers don't actively keep this in mind while they're coding, but they do need to have a basic understanding of this process. Now that you have the same general idea, let's look at how a programmer goes about stringing those instruction sets together.

What is Software Development?

In the broadest sense, software development includes all aspects of the process of developing software. That includes the conception, the research into how to do it, the initial design on paper, the prototyping of the system, the maintenance of the system once it's built, and the management of the actual processes involved. However, at the end of the day, it's mainly about the programming of the system you seek to build.

So, What Is Programming Anyway?

At the core of all software development, whether you're creating a simple website or a multimillion-dollar game, is programming. Programming is the process of creating a list of instructions for the computer to execute. A programming language allows the programmer to turn words and numbers into something the computer can understand.

In the lousy old days of computers, programmers created software by directly manipulating the instructions on the processor. They wrote sets of instructions for the processor to run, including the order they wanted them to run in to get their results. Using binary code, programmers communicated with the machine to tell it which instructions to run against what variables—they would pass the computer some numbers and tell it to add, subtract, print, or do whatever they wanted it to do. This took an incredibly long time, and, as a modern-day programmer, is pretty difficult to even comprehend.

In my grandfather's day, programmers created software using assembly code. Assembly is a human-readable language, where the programmer can refer to an instruction by a number and pass it human-readable values, like 1, 2, 3, 4, "add 1 and 2," "Hello, my name is Bob," and so on. Assembly is still used today when advanced programmers need to code something that operates at a very basic level (like running your graphics card drivers) or something that needs to be very fast. Most modern coders have likely never even seen assembly languages, and most likely will never need to see it.

Programmers today write software in higher-level languages. The programming tools they use translate more human-readable code into lower-level code that the computer can ultimately understand. For example, a modern programmer working with C++ (one of the world's most popular programming languages) could type:

```
a = 1;
b = 2;
c = a + b;
```

And because 1 plus 2 is equal to 3, "c" would be 3.

Today, most modern programming languages have tons of features included to make the developer's life much easier. For example, Visual Basic (a very popular language for Windows programs) could give you the name of the month with a simple command:

```
response.write (MonthName (8))
Would output: August
```

Clearly, that's a lot less work than writing the code needed to have an index that associates the names of the months, along with the number of months in a year (which you would have to do if you didn't have that clever function).

Modern languages are full of these kinds of nice shortcuts. Some specialized languages have functions to do very complicated tasks with one line of code. For example, the Wolfram programming language can do things as complex as pulling out all the education details from your friends' Facebook pages with one or two lines of code.

How does all that computer code get turned into something the CPU can understand? If you remember, computers only understand binary code (or machine language), and not lines of text in a programmer's toolkit. Well, all modern programming languages are either "compiled" or "interpreted." These are the two methods of translating lines of programming into machine language.

Compiled Languages

Compiled languages use something, surprisingly, called a compiler. A compiler is a special program that converts the programmer's code to a lower-level format the computer can understand. In other words, it will take something like "1 + 2 =?" and convert it to a set of instructions and binary data that the computer can run. Compiled programs are the kind you are likely most familiar with in your day-to-day dealings with a computer. Microsoft Word, Adobe Photoshop, and most of the apps on your smartphone are all examples of compiled software. They are complex and fast. Here are some compiled languages you may encounter:

- C++
- Objective-C
- Swift
- Go
- Java
- Delphi
- Fortran
- COBOL

Interpreted Languages

Interpreted languages, on the other hand, work a little differently. In the most basic sense, programs written in interpreted languages run inside other programs that "interpret" what to do with that code. Websites are the most common example of this today. When a programmer codes up a website, they don't compile their code. The web browser itself has a set of rules that takes the given code and then does what the code tells it to do. Your web browser is a good example of a compiled program, and the website inside it is an interpreted program. Here are some interpreted languages you may encounter:

- Ruby
- PHP
- Perl

- Python
- JavaScript

There are hundreds of programming languages out there. Web applications and mobile apps can be built with both interpreted and compiled languages, so don't worry too much about which is better; your development team should be able to help you decide on the right tool for the job. At the end of the day, all programming languages are doing the same thing: providing a number of shortcuts for software engineers to string together recipes of instruction sets to create something useful.

CHAPTER 3
TYPES OF APPLICATIONS

Software comes in all shapes and sizes. In this section, we'll explore the broad categories of software at play for commercial, marketing, and internal use. We'll take a look at the four main types of development you or your organization can engage in to make life for your customers or employees easier and more profitable.

These four categories are:

1. Traditional software
2. Mobile applications
3. API plays
4. Web applications

Traditional Software

For the better part of the last century, software came in only one or two flavors. There was software that ran on a single machine (e.g., Microsoft Word, Windows, printer drivers) and software that ran over a network (e.g., your real estate agent's property listing software, airline booking software, and ATMs).

As hard as it is for someone born after 1980 to imagine, for most of computing history, there was no Internet! One system would require software written in one programming language. The software created needed to reach the machine it was going to run on via a disk, a tape, or (I shudder to even consider it) a stack of punch cards.

Things have changed. My laptop doesn't even have a CD or DVD drive. Even if I wanted to, I couldn't install something I bought in a box at the store. These days, most software delivery occurs through the web or an app store like Google Play or Apple iTunes. But that doesn't mean traditional software is dead.

By "traditional" software, what I really mean is "desktop software." We (us nerds) used to refer to this as "application software," because it allowed the computer to perform useful tasks through an application, but that term is now pretty muddied. Desktop software is still very much alive and well. The word processor I'm using right now is a good example. You might use Microsoft Word or Apple Pages or something similar, but the idea is the same. You either buy a disk of some sort or download the software from the web, then install it on your computer.

Installing is something most of us have done a hundred times, but what does it really mean? Well, when traditional software packages are finally delivered to you, they are not normally ready to execute right out of the box. Usually, they have to copy themselves from a compressed file or disk onto your computer's hard drive. From there, they have to set up all sorts of permissions with the operating system so that the computer permits them to run. Finally, they copy specialized files throughout your computer's system so it can handle certain tasks.

Take, for example, installing Adobe Photoshop onto a Windows computer. When you get the Photoshop disk out of the box and stick it in your disk drive, it automatically loads another smaller program to handle the installation. The software copies itself to the computer's hard drive. It also copies files for things like fonts and specialized tools into locations other than where the main program will live. It sounds more complicated than it should be (and probably is), but this is the reason you have to uninstall programs when you no longer need them.

Uninstalling runs a program that remembers where all those other files are located within your computer and deletes them. If you decide to just delete the program from the folder you run the application from, it will mess up all sorts of things, resulting in errors every day and, most likely, headaches, high blood pressure, and expletive-laden speech syndrome.

These types of programs are almost always compiled programs. They are written in a specific language, for a specific operating system, and even for specific types of hardware. This means you can't go and buy the Windows version of Photoshop and install it on a Mac. To illustrate the point, the developers who work on Microsoft Office for Windows are a completely different set of folks than those who work on Microsoft Office for Mac.

If you want to build a traditional desktop application, you're going to have to spend a similar amount of money building it for one system as you spend on another. However, there are technologies out there that are cross-platform, meaning they work on multiple computer systems. Java and Adobe Air are both examples of programming systems that run on pretty much any computer. The catch is that your users will have to install a secondary piece of software, called a "platform," that allows that type of software to run on your specific system.

Pros of Traditional Software

Most traditional desktop applications are fast, plain, and simple. It makes sense that something written specifically for a certain computer is not going to have to worry about anything other than running well on that one system. This comes down to the compiler, that little program that turns human-readable code into something the computer understands. Compilers do lots of complicated things in this process, but they are designed to take advantage of the specific architecture of an individual system.

The other nice thing about traditional systems is that they always look and feel "native" to the user. By "native," I mean they fit the set of analogies and metaphors that a specific system employs to let their users interact with the computer.

Consider the menu bar on most modern applications. On a Windows machine, if you want to access the settings or preferences of a piece of software, you have to click "Edit," then "Preferences" in the menu bar. The menu bar is always attached to the window at which you're looking. Mac users are used to clicking on the name of the application that floats in the top left of their screen and then choosing "Preferences."

If this doesn't sound like a big deal to you, test it! Ask a user of another operating system to sit down at your computer and perform a simple task. Revel in their frustration as they yell something like, "Stupid Windows computers!" Or, "This is why I hate Macs." It not only inspires insight into the whole native software question, but it's fantastically entertaining.

Cons of Traditional Software

Large desktop applications are great. They handle most of the heavy lifting that people use computers for, like word processing, graphic design, and video editing. As you can imagine, building desktop software is quite challenging. The programmers required for this kind of work are fairly specialized and expensive. Your price goes up even more if you're publishing software for more than one operating system. Desktop applications are also much more difficult to distribute than online tools. When was the last time you bought anything that came with a CD to install?

As the web matures, many traditional offline applications are moving online. This created a lot of competition for the old giants in the industry. Google Drive handles most of the functions of Microsoft Office, and it's free. That must make life a lot harder for the folks at Microsoft, but all is not lost for these large software firms.

QuickBooks sells a desktop version as well as an online version. The online web version is subscription-based, starting at $15 a month forever—a much better bottom line than the $199 once-off price tag. The difference is that the types of technologies—and, as a result, the required programmers and their skills—need to change to keep up with the new paradigm. This will always be the case with technology.

Mobile Apps

If the web was the last big tech boomtown, mobile is the current exploding tech-megalopolis. There are now more smartphones sold each year than laptops and desktops combined. The amount of computing power in your pocket is growing much more quickly, by comparison, than the humble PC. The mind boggles at what's to come, and everyone wants to get in on the action.

Apple's iPhone really started it all. Like every major innovation, most people couldn't understand why they would even want one . . . until they had one. Now, you would have to pry my phone from my cold, dead hands.

The smartphone accomplished what the personal computer never really could; it made technology truly personal. My iPhone is now listed in every keys-wallet-sunglasses list I mentally check when leaving anywhere. I've left my phone at home once or twice and experienced something pretty close to separation anxiety.

While I might not be the average technology user, this isn't an uncommon experience, and for good reason. Having the power to be connected at all times may be a burden to some, but for most of us, it is freeing. You can now finally go to your kid's swim meet and still answer an email—or play Candy Crush if swim meets aren't your thing. Yes, it means you are not as engaged in what is going on around you, but could you even attend the swim meet if you had to be at your desk to receive an important email?

The biggest change is this ability to remain connected to the Internet at all times. As a result, the way we do business changed, too. These days, my phone wakes me up, and the calendar tells me where I need to be. Every meeting I sit in has someone with an iPad and Evernote. We then grab our phones and check Yelp ratings for where to have lunch. Google Maps tells me the quickest way to get back home, and I can log into my DVR on the way to make sure I'm recording the soccer game. After dinner, I read a book on my tablet or watch a movie on Netflix. There is almost no part of my (and millions of others') daily life that is not somewhat orchestrated through my mobile devices and the Internet.

Mobile devices coined the cute term "apps" to describe the applications that run on them. Apps are very similar to traditional desktop software, in that they are often, but not always, written for one platform at a time. The two biggest players in the market are Google with the Android operating system and Apple with iOS. Like desktop applications, apps are compiled software. You can't run an iPhone app on your Samsung Galaxy Android phone. As a developer, this has a lot of implications for deciding how you launch an app. Do you build your app for just one system? Do you build for both? Or do you launch on one platform first, and then start working on the other? We'll explore these questions in more detail in a later chapter. They all require different technical skills, budgeting considerations, design decisions, and launch strategies.

Both the Google Play Store (the main Android app marketplace) and the Apple App Store are above the one million mark in number of available apps. This is pretty staggering, considering the iPhone only came onto the scene halfway through 2007.

Now, while a large chunk of these apps are pretty terrible, it's amazing to consider the explosion in a new market. There is no reason, in almost any business, not to get in on this action. I don't say that because I'm an app developer and looking for work (we're booked pretty solid!), but because of the current customer expectations that this growth created.

One of the first things people do when they start engaging with a new company, or after they've purchased a new product or service, is check their app store to see if there is a nice, easy way to interact with their new purchase or service company. Whether you're a multinational bank or a local cleaning service, customers expect to be able to send you a message, check their bill, schedule an appointment, or access whatever they need, right now. From their pocket.

Did I mention I want it now?

If your company doesn't have a mobile strategy in place already, you're already behind.

Pure API Plays

For most non-technical managers, the API play is perhaps the most esoteric of software strategies. API stands for "application programming interface." Basically, an API is just a common language for one software system to talk to another. For example, if you wanted to add a driving directions map tool to your restaurant ordering application, you wouldn't create an entire map system from scratch. That would take years. Instead, your developers would probably grab something like the Google Maps API and use Google's custom instructions to integrate their very cool tools into your application.

Google Maps is written in the compiled language C++ (read as "C plus, plus"). C++ is a fairly complicated language to master, so you wouldn't want to have to ask Google Maps questions in C++. In fact, Google doesn't let programmers directly interact with code—that could lead to all sorts of problems! Instead, they offer an API for a number of languages. This permits programmers to ask questions of their system, and Google returns the answers in a way that can be understood by the system the developer is working on.

So, if a programmer wanted to create a map and center it on latitude and longitude of -34.397, 150.644, they could use JavaScript to ask for that particular map, which would look something like this:

```
center: new google.maps.LatLng (-34.397, 150.644)
```

Through their API, Google allots other programmers access to their tools, and without worrying about giving away their valuable code. Even better, they can charge companies who make lots of requests to their service. Now, most organizations don't make anywhere near the number of requests to Google's APIs to get charged for it, so if you wanted to use it, it would most likely be free. At full scale, a company making 100,000 requests a day is going to get billed $75 a day. That might sound like a high threshold, but thousands of companies make many more requests to that service every day.

This leads us to APIs as an actual business model. There are now hundreds of thousands of organizations whose sole profit model operates this way, and the types of APIs out there are too numerous to count. Here are just a few of the most popular APIs you'll find in many applications out there:

- **Facebook:** The Facebook "Graph API," as it's called, offers hundreds of different features, including the ability for applications to post to Facebook, list friends, and use Facebook as the login system for other websites and apps.

- **Yelp:** The Yelp API gives developers access to their reviews and businesses.

- **Foursquare:** The Foursquare API lets programmers access Foursquare's massive database of locations around the globe, including the ability to create check-ins and see who checked in to those locations.

- **Dropbox:** Dropbox has a great API wherein applications can store and share their data on Dropbox.

- **Stripe:** Stripe has one of the most useful credit card payment gateway APIs out there. Stripe will store your users' credit cards and process them, without exposing you or your organization to the risk involved in handling your users' sensitive credit card information.

- **MailChimp:** MailChimp offers an API with the ability for systems to use MailChimp's email services to manage large contact lists and automatically send bulk email to thousands of users.

- **Twilio:** Twilio's API gives developers the power to integrate text messaging (SMS) and phone integration into their apps. This means you can easily create a phone call-in system, send users SMS updates, automatically call their phones to confirm purchases, and utilize thousands of other functions.

And there are millions of other great APIs out there for everything from the commonplace to the extremely obscure. For example, the Pirate Translation API allows you to convert any sentence into Pirate talk! Aye not 't sure why ye mateys would use this, but fun it may be!

The idea behind creating an API for your business relies on the technology underlying the API. If your company has certain in-house IT systems they want to give outside developers access to but not the general public, creating an API serves as the perfect solution. In most cases, APIs have added bonuses to existing web or mobile applications.

If you're developing a web application that lets people custom-print T-shirts, it would be nice to have an API you could offer to developers who also want to build a T-shirt printing website. They could take your API and easily make their own site. All their orders would come through your system (with, presumably, some sort of markup). Everyone wins in this situation. If the API does something more complex, and the power is in the information rather than an actual product or service, you could charge for the API alone.

Imagine you were building a system that analyzed the prices of businesses sold in mergers and acquisitions. The system made assumptions as to what the state of the market would look like in that particular sector. That system might be advantageous to other companies in market segments other than yours. A commercial API would then allow you to charge developers to access your technology.

This is the main difference between APIs and other types of software. While most applications target consumers or businesses with their offerings, APIs are mainly sold to other developers (and their organizations). Traditionally, API prices depend on "API calls," or the number of requests made to the provider's server.

"Server" is one of those tech words you've probably heard a million times without really understanding what one is. It might sound like a complicated and scary thing, but it's just a computer hooked up to the Internet twenty-four hours a day. Each server receives a special address, called an IP address, which tells the world its location. For example, if you type 63.96.4.59 into your web browser, you'll get Google.com. Companies like GoDaddy (famous for their over-the-top TV commercials) and Network Solutions translate these IP addresses into names we can remember, like cnn.com or bing.com.

If your system made a million API calls to your provider each month and they charged $0.001 each, you would get a bill for $1,000 at the end of the month. There are a number of other less common pricing structures for APIs, including daily and monthly access flat fees, charging based on the number of users using the API (rather than the total number of requests), and charging for the amount of data sent over the API (for example, getting charged per megabyte uploaded by your users).

Most software systems these days have some sort of interaction with the Internet. This means you build some sort of server application no one in the outside world actually sees and then you build a frontend application (e.g., a website or mobile app). A frontend is, as the name implies, the face of an application. This is the (hopefully) pretty part users interact with.

Next, link the two together. For example, if you build a mobile app that requires your users to log in and create accounts, you store those accounts on the Internet, somewhere on a server. Rather than accessing the server directly (which can create security issues), developers will build an API for their own systems, which the app uses to communicate with the server. This lets developers build the backend server in one technology and the mobile app in another. The **backend** comprises everything that makes the frontend work—the nitty-gritty stuff that goes on behind-the-scenes. While a very common practice, the APIs don't ordinarily get published to the outside world; they are only used for the internal communication of the system itself.

Web Applications

The Internet is by no means new, though most people didn't start using it until the invention of the World Wide Web. Email and the web have changed the way we work, play, and interact with others more than any technology since the printing press. That may sound hyperbolic, but no form of communication has spread as rapidly, and leveled the intellectual playing-field of the world, quite like it. College kids in their dorm rooms now plan on changing the world because they actually might be able to do it. People in the poorest parts of Africa learn to build wells and wind turbine generators from Google searches on solar-powered laptops and cheap cell phones. The Internet touches, in some way, every aspect of modern business.

The Internet is not just email and the web. There are hundreds of other services that run over the same networks. The average person is just not exposed to them. Everything from your ATM to your home video service is now completely reliant on the cables, microwave dishes, satellites, and cell towers spread across the globe. So, obviously, everyone wants a piece of it! The web is a shear innovation engine. Web development spawned an entire industry of programmers and engineers building applications delivered to you through a browser.

A web application is simply a program interpreted by a web browser. The code the programmer writes is not compiled, and can only run on one system, but it is left in its code form. The web browser (e.g., Internet Explorer, Firefox, Safari, or Chrome) looks at that code and works out what to do with it.

The browser itself is a compiled program. It knows what to do with the commands it receives from the web server.

Once your web browser receives the name you type in, it checks with the next computer on the network until it gets the IP address. It then follows the numbers, similar to the way you follow house, street, and zip code instructions to arrive at a physical address. Finally, when it reaches the address, the computer waiting at the other side has a program running on it that says, "Hey! I'm that-site-you-were-looking-for.com! Here is the website you wanted." Of course, it says this in a much less friendly way.

When the web was young and innocent, all it could understand was a very simple programming language called HyperText Markup Language, or HTML. We called the language "HyperText" because it consisted of a set of links that would jump you from one document to another and "Markup" because markup languages are very simple methods of formatting documents with text. For example, if you were reading this on a web browser, I could write:

```
<i>italic text goes here</i>
```

The "italic text goes here" would look like this: *italic text goes here*. It's not exactly the type of programming that requires math and logic training, but it was a good start.

HTML also understands how to do basic layout functions like making tables, displaying images, and, obviously, making links. This was the state of the web for a long time.

My first real job was building these types of websites for a software company on the cutting edge back in 1995. I was a kid, and the company was growing so fast I didn't even have a desk or a computer! I sat on the floor in the new Web Design Department, using the CEO's laptop, and built sites for companies for hundreds of thousands of dollars. If only they knew that a sixteen-year-old, getting leg cramps from hours of floor sitting, was doing the work!

But those days didn't last long. People started teaching the web, web browsers, and new tricks pretty quickly. Serving up text documents with clickable links was useful if you wanted to tell the world about your company, but up-to-date information was near impossible. Having anything short of pictures and paragraphs was inconceivably expensive. The web needed real intelligence and logic. Enter scripting languages!

There are two fundamental types of programming for the web. These are the basis of everything from a one-page do-it-yourself website to the largest web applications out there.

Server-Side Scripting

The first big trick that programmers learned to make the web more interesting was server-side programming. This is pretty much what it sounds like. A programmer would write a traditional program that would run on a server. The difference between a server-side program and a desktop program is that the server program looks for inputs from a network—in most cases, the Internet.

The website takes some data from its very limited environment of the browser and HTML, and it passes that information to the program that is running on the server. This is what happens every time you fill out a form on the web today. The web browser doesn't know what to do with that data, but the server says, "Oh, you want to send your request to get a reservation for our restaurant to the manager? Ok, not a problem. I'll take that data and email it to her for you. Ok, I'm done! I'll send the browser back a message saying 'thank you for your reservation'. Thanks so much!"

In this way, your very limited web browser can do, or at least appear to do, so much more. You're putting the frontend of a much more complex program on the web. That program can interact with databases or complicated mathematical algorithms or whatever the programmer dreams up. Now, while that's a great start, it still leaves the problem of a very limited, and stupid, web browser. It can't do anything cool, like check to make sure your phone number has nine digits, without the help of a server.

Think about it this way. If you type your phone number into a form, and we want to check to see if it's a proper U.S. phone number (i.e., xxx-xxx-xxxx), we have to send that to the server, wait for the program to do the check and send back the answer, and display it on the screen. You've probably experienced this before after filling out a form, hitting the "submit" button, and then seeing an error message. This is pretty annoying for a modern web user. Why did it have to reload the screen? Why couldn't it have told me I did something it didn't like before I hit the submit button? Well, the reason is that the submit sends that information to the server. The server then sends the response back to the browser (called "the client" in this case). But this is steadily becoming a thing of the past in many well-written systems thanks to client-side scripting.

Client-Side Scripting

Browsers grew much smarter in the last few years. No longer are they just able to understand HTML, but they can interpret lots of other technologies. For many simple and now increasingly complex tasks, browsers can handle the whole load themselves. These days, if you submit a form and it comes back telling you your phone number didn't have enough numbers, you can be sure it's either an old site or run by very busy programmers taking shortcuts.

The most common client-side scripting language is JavaScript. JavaScript has nothing to do with the server-side or the traditional programming language called Java. This confuses most people when they talk to nerds who use both technologies. Even veteran IT managers often don't know the difference. This is probably because marketing folks coined the term JavaScript to get some of the name recognition of the older, more established technology. But I digress!

JavaScript and its other client-side brethren allow programmers to run programs right in your browsers. You don't have to wait for the server to work out what to do with that phone number you typed in—the website loads a program into the browser and tells it to generate a red box when you don't input the number correctly. This is one small example of what the newest client-side tools can do, but it illustrates the point. These days, entire games, full word processors, graphical editing suites, and even databases run on their own in the humble web browser.

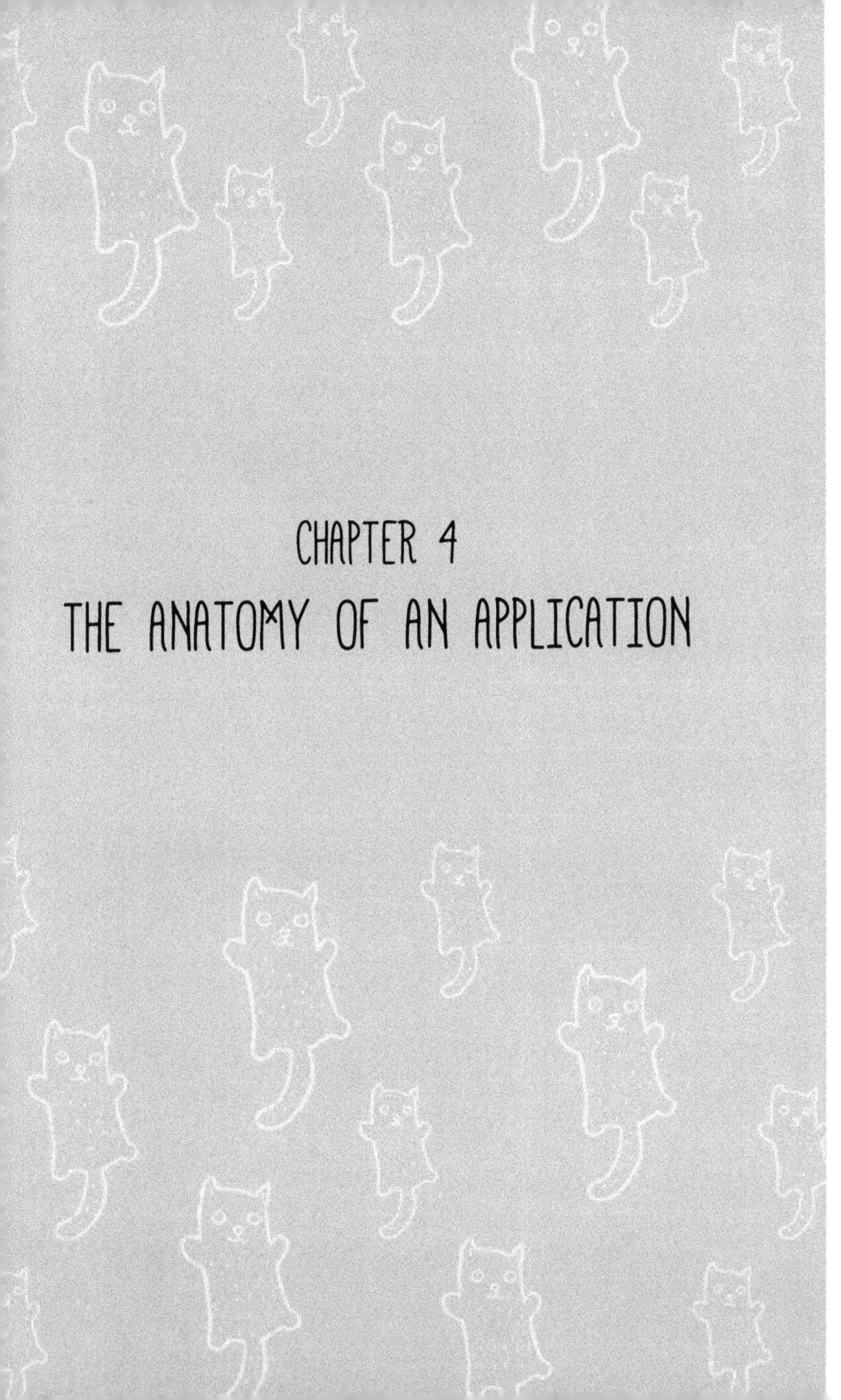

CHAPTER 4
THE ANATOMY OF AN APPLICATION

In this section, we'll look at the basic building blocks of any software project. Most of these should seem reasonably obvious, but it doesn't hurt to go over them and make sure you know the terminology that might get thrown out at meetings with nerds. Engineers like specifics. If you can be specific about anything, they'll probably respect you more and be easier to work with!

Applications of almost any kind follow only a few simple patterns:

1. They include a way for a user to interact with them—a "frontend" or "view(s)."

2. They have a place to store data the user will interact with or generate—normally a "database."

3. As with most frontends, there is a "backend!" This refers to the stuff behind the scenes that makes an application work. It can also refer to a set of tools or systems through which an administrator can interact with the logic of the application or the data.

4. The newest addition to these lists is "third-party integrations." Third-party integrations include anything you don't build, but rather plug into your system to extend its functionality. A Share on Facebook button or Google Maps in your app are good examples.

5. Finally, there will be some connective tissue between the systems. The broadest term for this is "middleware," a term so broad we could write an entire series of books on this topic alone.

These basic building blocks are what most modern-day Internet-connected applications rely on. Yes, applications exist that don't have user-facing frontends and only talk to other computers. Yes, there are applications that don't use databases or store any data at all. For our discussion, we'll focus on the generalities of modern web and mobile applications. So, let's get started!

The Stack

As more and more software delivery occurs via the web or mobile technologies, the infrastructure that delivers these applications becomes increasingly vital. Programmers write traditional software in a programming language suitable for the intended platform. This software is then shipped to the user on a CD or via a download. It can run without the user connecting to the Internet.

Web applications, however, are Internet-only, and many mobile applications require a network connection to function. Why? They need to "phone home" to a server somewhere that does a lot of the heavy lifting. Very few software products these days rely on a single technology. There are different languages, software packages, and frameworks that make up the anatomy of a modern application. We call this combination of technologies that give life to a system the "stack."

In the case of a web application (everything from a simple blog to an enterprise management system), the software needs something to "live" on. Enter the server!

As mentioned in the previous section, a server is a computer hooked up to an always-on Internet connection and some server software. The special software listens to the network all day long, and when it obtains a certain command, it wakes up, does some work, and then responds back to the thing that sent the request.

The two most popular server software packages out there are the Apache web server and Microsoft's IIS (Internet Information Services). There are hundreds of other web servers in production, but these two account for around 70 percent of the market share.

Servers allow requests from the network to be answered in a number of ways. At their most basic function, they serve up simple HTML files to the user. Servers also "parse" scripting languages into commands that the server will run.

Nerd Out

We call the model of a device or browser connecting to a server the client-server model. The client in this case is the web browser or application asking for something, meaning whatever consumes the responses from the server. You'll often hear developers talking about "the client." Don't worry, they're not talking about you!

Developers refer to programming software that runs on the server-side as "backend programming," whereas programming created to run on the client side is "frontend programming."

Frontend Development

The Basics

Anything on which you use a keyboard, mouse, touch screen, etc., is technically a frontend of some sort.

In the olden days (anything before 1999), "development" often meant "backend programming." As the complexity of modern applications increases, so too does the importance of frontend development. Frontends consist of two main components:

1. Design: How the application looks and feels—the graphic assets, logos, buttons, forms, and so on.

2. Functionality: How screens flow from one to another, how information pops up, and anything that changes in real time when a user interacts with it.

As you might imagine, the design process is similar to traditional print design in which a designer uses graphics programs to "draw" pictures.

The functionality process is more of an exercise in programming, using programming languages to make the designs come to life. Even so, the line between the two processes blurs more and more each day. There are plenty of cool tools traditional designers can use to make their designs come to life without having to be hardcore coders, and old-school programmers have a plethora of libraries and tools they can import into their code to make their bland functionality look more friendly. At our company, we don't have any so-called graphic designers. We have "frontend engineers," which is a fancy title for people who can make pretty pictures and turn them into usable interfaces with code.

Graphic Design

Graphic design is an artistic pursuit. While the title "graphic designer" could refer to everyone from Andy Warhol to 3D modelers working on next-generation video games, it means the same thing: visual communication.

The skills required to be a graphic designer vary radically depending on their medium and focus. We're going to stay away from the visual arts side of this topic because I'm not a qualified art critic, but let's delve into the design side of applications.

When it comes to the world of digital, there is a lot of crossover in terms of skills from print designers to software frontend designers. This is due to the ubiquity of one software application: Adobe Photoshop. Photoshop is easily the most-used software program out there for designers. Whether you're making an ad for a newspaper, touching up a photograph, or designing the texture of a wall in a virtual reality system, odds are, you're using Photoshop. It's so dominant in all design industries that it's often used as a verb: "I'll just Photoshop it a bit, and it'll be awesome!"

Photoshop is a tool that lets designers manipulate pixels, the smallest controllable element on a screen. A modern screen has millions of little pixels, which can be lit up with a number of colors. With Photoshop, designers draw shapes, paint with virtual brushes, add text, and create a myriad of elements. If a designer is using Photoshop (or any other graphical tool) to design the look and feel of an application, they will often chop up their design into parts like buttons, backgrounds, banners, or whatever else, and send those off as separate files to the programmer. The programmer then places them on the screen they're building and wires them up to make them do interesting things.

Most professional web, mobile, or desktop application designers end up writing the code themselves to make the designs come to life. This code might include complex behaviors, like how a button changes when you move your mouse over it, or how it looks when you click on it. Next time you click a button on a website and it looks like it depresses, think about the frontend designer/developer who had to make three different images and write a ton of code just to make the button do that!

Nerd Out

There are three main types of graphic design. The first is "raster graphics." Raster graphics are a type of image built up with pixels. If you've ever zoomed into a photo enough that it becomes blocky, that is a raster graphic. Those blocks you see as you zoom in are pixels. Photoshop is a raster graphics editing application.

Second is "vector graphic" design. Vector graphics are images built with tools similar to those found in Photoshop that draw the images on-screen. Vector graphics do not manipulate pixels. Rather, they create mathematical formulas that describe things like the curve of a line or the gradient of color between two points. If you zoom in on a graphic and it never gets blocky, no matter how far you zoom, that's a vector image. We use vector graphics for things like logo design, where you might need to display the image on a phone screen or a billboard, and you don't want to lose any quality.

Finally, we have "3D" design. Think of a Pixar movie or most video games released in the last twenty years. Tools like Maya 3D or 3D Studio Max (both made by Autodesk) bring 3D graphics to life. These are similar to vector graphics, in that the designer uses lines and shapes to build skeletons of objects (like building the structure of a papier-mâché sculpture using chicken wire). They then "skin" the skeleton using textures made of raster graphics (often made in Photoshop). The computer renders the animation, or image, using clever algorithms to work out the light and shading set up by the designer to make the object look realistic.

UX vs. UI

Before we go any further, I do need to mention a technical sticking point that comes up time and time again. This is the difference between user experience design (UX) and user interface design (UI).

UX is the design of the way the system will work. This does not mean how it will look. Typically, UX design consists of a series of wireframes (more about these in the Planning chapter). UX is where you start your design process long before bringing a graphic designer into the picture. An example of UX design is a sketch on a piece of paper showing the general idea of a mobile app. It could have very rough representations of user login forms, showing a box for "username" and "password" with a "Log in" button. These would intentionally lack in design. Think of this as the blueprint for the application. UX design can often include actual working, interactive features. This is so users can test how the system works before the creation of any code or graphic design.

UI design, on the other hand, refers to the pretty pictures, or "skin," of the system. These files look exactly like the final product presentation. If you saw them on a screen, you might try to click on the buttons or start typing into one of the forms, but they're just pictures. These are the screens, most likely created in Photoshop, eventually cut up into the separate elements the developer will use to build the frontend.

These two terms are repeatedly thrown around in the same sentence in development meetings and can become a little confusing. In fact, I know a lot of programmers who use the terms interchangeably, even though they're not the same thing.

Programmers don't really care about the design of the product. At the end of the day, a lot of developers are only interested in making sure their code works and the system they're building is stable and secure. This is partly the fault of developers, for sticking to their silos, and partly the fault of traditional designers, who want to hang on to their skills and not learn new tools.

More and more, you're finding full-stack developers. This is a term for guys and gals trained in building both the backend and frontend of a system. They're either going to want to be hands-on in the design process or have a lot of input before they start coding.

The Development of Design

We've now got some great graphical assets showing how our website or application will look. How does that translate into a product that does something other than look nice? As always, with life as well as with computers, it depends.

The process laid out in the following pages focuses on developing applications for the web. You'll find that the principals are exactly the same for many technologies. Tens of thousands of mobile apps for Android and iOS are built in this way, as are interfaces for TVs, phone systems, and even some desktop applications. This process is the holy trinity of development using the technologies HTML, CSS, and JavaScript.

While these technologies aren't used to create most desktop technologies and native mobile applications (more on that later), the fundamentals are similar. Those other systems often have their own priority technologies, methods, and even applications you must use in order to build their interfaces. By contrast, the technology "stack" I'm about to introduce you to is universal.

Because these are the foundational languages you interact with in day-to-day life, I'll go a little more in-depth with them.

HTML

HyperText Markup Language (HTML) is, well, a markup language. (Not very helpful, I know, but bear with me.) What this means is that it takes existing content (text, images, links, tables of data, and so on) and adds little tags around it to give it form. HTML is a very basic language mainly used to lay out the areas of a frontend, with broad strokes.

For example, read through the following code:

```
1   <html>
2       <head>
3           <title>This is an example of HTML</title>
4
5       </head>
6       <body>
7           <h1>This is a heading!</h1>
8           <img src="http://upload.wikimedia.org/wikipedia/commons/thumb/6/61/
            HTML5_logo_and_wordmark.svg/2000px-HTML5_logo_and_wordmark.svg.png" width="300px">
9           <br>
10          <!--This is a comment! it's invisible to the browser, and is only for people to read. That <br> tag
            there is a a line break-->
11          <h3>And this is a smaller heading!</h3>
12          <p>
13              This is a paragraph of text. Those little tags you see around my text tell the browser how to
                interpret them.
14              For example, if I wanted to make a word bold, I would use the strong tag, <strong>like this</
                strong>. As you can see, the period is not bolded, because it's outside of the tag.
15              <br>
16
17
18              My image about (that img src= thing) tells the browser to display an image for me. In this
                case, the HTML5 logo that I grabbed off
19              <a href="http://www.wikipedia.com">Wikipedia</a>. And just for fun, I've added that link tag (a
                href=) for the Wikipedia link. If you type out this bit of code into most text editors, and
                save it as a .html file, you will be able to open it in a web browser, and it'll work!
20          </p>
21      </body>
22  </html>
```

So, if we run this code in a browser, it'll look like this:

This is a heading!

And this is a smaller heading!

This is a paragraph of text. Those little tags you see around my text tell the browser how to interpret them. For example, if I wanted to make a word bold, I would use the strong tag, **like this**. As you can see, the period is not bolded, because it's outside of the tag.
My image about (that img src= thing) tells the browser to display an image for me. In this case, the HTML5 logo that I grabbed off Wikipedia. And just for fun, I've added that link tag (a href=) for the Wikipedia link. If you type out this bit of code into most text editors, and save it as a .html file, you will be able to open it in a web browser, and it'll work!

If you're thinking to yourself, "That looks easy! But it does look a little like the web in 1996," you'd be right. The web in 1996 didn't have much else going for it other than HTML. It was awful for a long time. Trust me, I was there! Fortunately, things improved a bit.

CSS

By the middle of the 2000s, CSS, or Cascading Style Sheets, started to give HTML a little life. CSS is another very simple language similar to HTML, but it focuses completely on how something looks, as opposed to HTML's focus on what is displayed. CSS focuses on defining the way tags in HTML look. There are hundreds of tags in HTML, and CSS lets you define their color, size, position on the screen, padding around them, and so on. CSS3 (the newest version of CSS) can even animate tags and make them behave differently depending on screen size.

To illustrate, let's take that same HTML code and add some CSS to it. As you can see, I am defining things like the font size and the background colors of the tags in my HTML to make them more interesting. As an aside, this is not great CSS code. It's just there to give you an idea.

```css
html {
    font-size: 1em;
    line-height: 1.4;
    font-family: sans-serif;
}
body {
    margin: 15;
    background-color: ##d1d1d1;
}
h1 {
    font-size: 2em;
    margin: 0.67em 0;
    color: orange;
    margin-left: 330px;
}
h3 {
    font-size: 1.17em;
    margin: 1em 0;
}
strong {
    font-weight: bold;
}
img {
    float: left;
}
p {
    background-color: #92d5ff;
    padding: 7px;
}
```

I then use one line of code to link the two files together:

```html
<html>
    <head>
        <title>This is an example of HTML</title>
            <link rel="stylesheet" type="text/css" href="example.css">
    </head>
```

Then I refresh my browser, and. . . .

This is a heading!

HTML

And this is a smaller heading!

This is a paragraph of text. Those little tags you see around my text tell the browser how to interpret them. For example, if I wanted to make a word bold, I would use the strong tag, **like this.** As you can see, the period is not bolded, because it's outside of the tag.

My image about (that img src= thing) tells the browser to display an image for me. In this case, the HTML5 logo that I grabbed off Wikipedia. And just for fun, I've added that link tag (a href=) for the Wikipedia link. If you type out this bit of code into most text editors, and save it as a .html file, you will be able to open it in a web browser, and it'll work!

Okay, I know it's not the greatest design in the world, but it looks a lot better than what we had in the first example!

With HTML and CSS, we can turn our design ideas into reality. What we have at this point is the framework for creating beautiful documents that link to each other and can have some minor functionality like glowing buttons or basic animation. What we don't have is any logic—there is no ability in HTML and CSS to do things like evaluate a mathematical equation or take in data from a user, do something with it, and return a result. This is okay because we're talking about frontend development here and not backend development.

Nevertheless, user interfaces grow increasingly complicated. Users now expect menus to expand and contract or buttons to perform tasks immediately. Remember that for most server-side programming languages to work, they need to send their data to the server, then have the server send back the response. Think back to my earlier example of hitting a submit button on a form, only to reach a page with an error saying your phone number was incorrect. Enter JavaScript!

JavaScript

JavaScript has a long, strange history, but it looks like it's here to stay. Most programmers pooh-poohed JavaScript for years, partly because it wasn't very complicated or powerful (it was originally targeted to hobbyist programmers), and partly because no one really knew how to do anything valuable with it.

With JavaScript, all we do is add one line of code to our HTML, which creates a button.

```
18
19        <button onclick="myFunction()">Click Me!</button>
20
```

ML

And this is a smaller heading!

This is a paragraph of text. Those little
my text tell the browser how to interpre
if I wanted to make a word bold, I woul
like this. As you can see, the period is
it's outside of the tag.

Click Me!

My image about (that img src= thing) te
display an image for me. In this case, t
grabbed off Wikipedia. And just for fun,

JavaScript will then run a "function," which I creatively called "myFunction." The function tells the browser to pop up an "alert box." I'm sure you've seen these a thousand annoying times! This time, I'm giving the alert box a harmless message.

```
1    <script>
2        function myFunction() {
3            alert("Well done! You clicked me!");
4        }
5    </script>
```

Functions are little, self-contained pieces of code that a programmer strings together to make more complicated programs. A programming language comes with hundreds of built-in functions and lets you create your own.

With these few extra lines of code, I can click that button, and JavaScript automatically tells the browser to launch the little alert box.

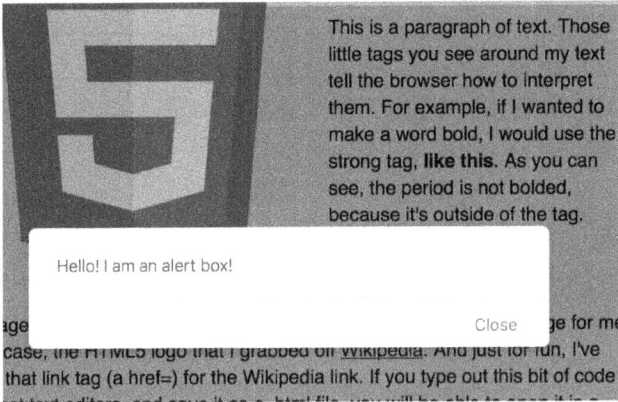

I've uploaded all of the code examples to https://www.sourcetoad.com/herding-cats/ if you would like to play with them.

JavaScript then allows us to create real interaction with our frontends. Of course, that's not all JavaScript can do. No longer is it relegated to the status of a second-rate programming language. Entire applications come to life with JavaScript. It runs inside web browsers, mobile phones, and PDF documents.

JavaScript has become a ubiquitous, legitimate language. This sometimes still seems very strange to older developers, because for much of its life it was not much more than a novelty. Yet the growth of the web and mobile provided the perfect nursery in which JavaScript could grow.

The Backend

Most web and mobile applications have some sort of administrative dashboard where the owner of the application performs certain tasks like creating new users, editing content, or simply viewing reports and statistics. This is often called the backend. Once again, we end up with a term that means different things to different people.

To most developers, backend means something a little abstract. Programmers often think of the backend as the underlying software with which the user (or client) does not interact. The scripts and programming do the work, not necessarily the interface that administrative users utilize. In fact, a software engineer would most likely refer to that as a frontend!

Time for a very simple code example. We're going to do some frontend and backend development here. This is a little more complicated than our frontend example in the last section, but most of this should make sense to you. We're going to build a simple calculator that only performs addition. Let's start with the frontend. This should look somewhat familiar.

```
<html>
<Title>Let's Add Two Numbers</title>
<body>
    <form action="lets_add.php" method="post">
        Enter a number: <input name="number1"
        type="text" />
        <br>
        Enter another number: <input
        name="number2" type="text" />
        <input type="submit" />
    </form>
</body>
</html>
```

I named this file "add.php." What we have here is some HTML code that presents the user with two input boxes. Surrounding our code are <form> tags. These tags "post" (or send) the output to a file called "lets_add.php."

In our web browser, the code looks like this.

Enter a number: 4
Enter another number: 2   Submit Query

It's not very exciting, I know, but it's only half the battle. I put the 4 and 2 in those boxes by hand because we need something to calculate.

Okay, on to the backend! I'll be using PHP for this example because it's the most common web scripting language (and I know it the best). Take a look through the code below:

```
<html>
<head>
<Title>Let's Add Two Numbers</title>
</head>
<body>
    <?php
        $number1 = $_POST['number1'];
        $number2 = $_POST['number2'];
        $answer = $number1 + $number2;
        echo "When we add the two numbers we get:".
        $answer;
    ?>
</body>
</html>
```

You may have noticed that this looks a lot like HTML. Indeed, all those <head> and <body> tags are HTML. Scripting languages live alongside other languages all the time. Let's look take a closer look at our code.

The first thing we do is tell the server that this is not HTML by inserting that <?php> tag. This alerts the server that it needs to take over and do something. In PHP, we use the dollar sign ($) to indicate a variable, so we tell it to create two variables ($number1 and $number2) and assign them the value from what was "posted" from the previous file. We then create another variable called $answer and assign it the value of the sum of the two numbers. Finally, we "echo" or print out some text, followed by the value of $answer.

When you hit that "Submit" button on add.php, it sends the data to lets_add. php and prints out the answer, which looks like this.

> ## When we add the two numbers we get: 6

Okay, so this is pretty close to a useless program. If you have access to a web server, you probably have access to a calculator! Yet it does show how an interpreted, server-side application works. The server here is doing the math, not the client.

You can't just type this code into a text file and open it in your browser like you could with the frontend example in the previous section. That's because you're not running a server. The server is what interprets our code and sends it back to the client. If we wrote the same code in JavaScript, we wouldn't need a web server, because JavaScript is interpreted on the client side.

So, why would we ever use server-side languages if we can do the same thing on the client-side?

Up until very recently, JavaScript couldn't do as much as server-side languages, so we had to use other tools. The general rule of thumb was to use the server-side tools to interact with items that live on the server, like databases, files, and other programs. We used JavaScript for objects that lived on the client-side, like text boxes, buttons, and animations. This line is now blurring. JavaScript will not replace the rich ecosystem of languages and available tools out there anytime soon, but a number of recent developments have made it a real possibility.

A Brief Word About Backend JavaScript

While there are a number of platforms a developer can utilize to build traditional desktop applications, web apps, and mobile applications, there is a relatively new kid on the block taking the software world by storm, Node.js.

Node.js, or simply "Node," is a JavaScript framework that allows developers to use JavaScript to do a lot of the things that a good old-fashioned server-side scripting language (like PHP, Ruby, or .NET) can do. I won't go into the nitty-gritty regarding Node, but suffice it to say that JavaScript developers, traditionally only frontend people, can now build full applications, access databases, send commands to the server, and do almost everything you once needed a backend developer for.

There are some cases where this makes a lot of sense, but also many cases where it does not. If your developer says she wants to use Node to build your system, ask her for a few good reasons. If you feel like the answer is "because it's cool," be cautious. I personally love Node, and many of my developers are big fans, but we have run into situations where it caused more problems than it solved. Node is a newer technology, and it will mature over time. But as with all tools, there is a right time and a wrong time to use it.

Databases

We have now discussed:

- The server, which listens and responds to requests
- The frontend, or the code that gives a user something to interact with
- The backend, or the underlying code that does the work on the server

These are all the components you need to create an effective, human-facing application. Most of the time, you're going to want that application to store and retrieve data in some way. For that, you need a database management system.

Database management systems (DBMSs) are the underlying software that give data some structure and a place to live. It's unlikely that you'll ever hear anyone say "DBMS," but if you do, that's what they're talking about. For brevity, I'll refer to database management systems simply as "databases."

You're probably familiar with the term "database." For most people, the term signifies a collection of data. If you hear a programmer use the word, they're talking about the collection—and, more likely, the system—that the database runs in.

Database systems themselves are software programs that allow the programmer (or database administrator) to define how data resides inside the database. They have a number of rules and commands as to how to insert, retrieve, edit, and delete that data. Let's look at an example of an application you might build in the real world to illustrate how a database works alongside code.

Imagine an application designed to track traveling salespeople. The sales staff have smartphones that include an app showing them all their appointments and permitting them to enter notes about how their meetings with potential clients went. To make it more interesting, we'll say the app also logs the GPS coordinates of the salesperson throughout the day so that the sales manager can make sure they're doing their rounds in a timely fashion.

We know the app has a frontend, through which the sales reps interact with the application. We also know that there is a bunch of backend code on the app to handle the logic of the app itself. For the app to prove beneficial to everyone involved, it needs to talk to the backend code on a server somewhere (where the sales manager can set the appointments, add new sales staff, and get reports on employees' movements). We could store all that data in plain old text files, or as HTML, or any number of formats. The problem? It's really difficult, and slow, to have a computer search through mountains of files for specific information. A database solves that problem.

In a database, we get to define the structure of the data. In this example, the programmer might determine a list of customers, a list of salespeople, a list of GPS coordinates tagged with a salesperson's employee ID, and so on. Once they define this structure, the programmer can use their code to ask ("query" is the term most often used) the database questions about any data it contains. The database is able to return a result or a set of results. Databases excel at doing this quickly—and somewhat intelligently.

There are a number of different ways to plan a database. The most common is what's called a "relational database." This is a way of defining your data structure so that there are meaningful connections in various data categories. We split relational databases into "tables." Think rows and columns. In this example, the "Salespeople" table might have columns named "Employee ID," "First Name," "Last Name," "Territory," "Phone Number," "Email Address," and so on. Think of it as a Microsoft Excel spreadsheet. In fact, many people use Excel as a database for things like this.

Now, what would happen if we now want to store a time and date stamp along with a GPS coordinate with that sales rep? Well, we could add another column called "GPS" or something similar, but that would only be one location. Theoretically, we could just make more columns, like "GPS location 1," "GPS location 2," and so on, but when you have the app recording every GPS location of the salesperson, every five minutes, every day, you would end up with thousands of columns!

A relational database gets around this by creating a separate table completely. Let's call this the GPS table. Its columns include "Latitude," "Longitude," "Time/Date," and "Employee ID." Now, every time the salesperson moves, the app adds a row to that table with the updated data and their unique identifier: their employee ID number. As long as the database knows that the Employee ID in the Salesperson table is the same ID associated with each GPS coordinate, a programmer could query the database and ask something like, "Please give me Bob Smith's employee ID number." Once obtained, they could ask the GPS table to "show me all the GPS coordinates of employee #123 between Monday the twenty-fifth and Thursday the twenty-eighth. And oh, please order them by date, then time." The database would politely respond with a long list of GPS coordinates, nicely ordered by date and time. The programmer could then use these coordinates to draw a map of that salesperson's movements.

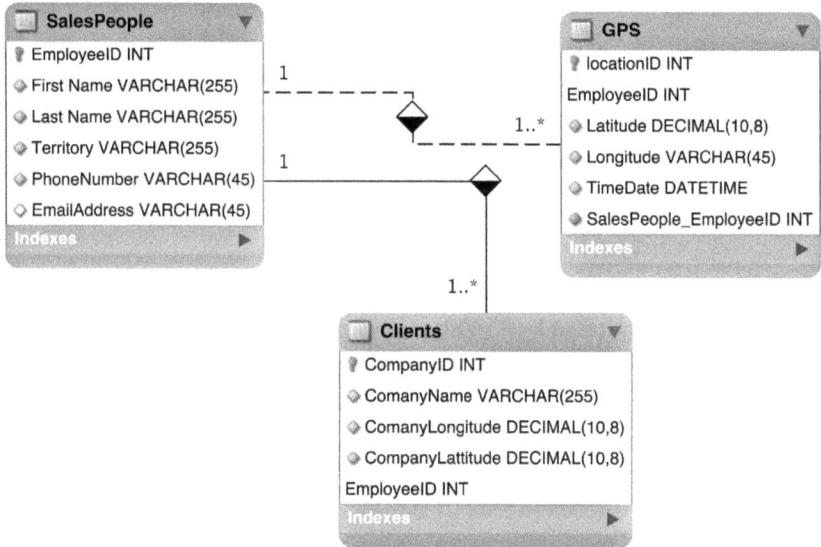

There are, as with everything in development processes, hundreds of choices in databases. We're going to look at the two most common types and a few examples.

Relational Databases

Relational databases are what I described in the previous section. They have tables (columns and rows of data) and relationships (links between certain columns in one table to a column in another table). This is the most common type of database in use today. If you've ever worked with a database, it was probably a relational database. We'll quickly go over some of the big players, so if a developer ever suggests one over the other, you'll have some idea of what they're trying to sell you.

Oracle

The 500-pound gorilla in the industry. Oracle is a monster database with a massive amount of features and tools. Oracle is blazingly fast and very scalable. The downside to Oracle is that it can get very expensive. This is because Oracle charges by the user or by the number of CPUs on the server running it, and modern servers can have multiple CPUs. Oracle licenses can run into the hundreds of thousands or even millions of dollars per year. There are complex pricing guides and rules and even fines if you abuse their license. As a result, Oracle is customarily reserved for enterprise-style companies.

Microsoft SQL Server

Microsoft's offering in the world of databases is also not cheap. It's reasonably affordable for small projects, but their enterprise-grade version costs $14,256 per core. My laptop has four cores in its CPU, which means I'm looking at $57,024 if I want to run it! Once again, this means that only larger companies shell out for Microsoft's top offering. To bring more developers onto their platform, Microsoft has begun to offer significantly cheaper options, which has paid off in large market share gains.

MySQL

MySQL is free, open-source, and massively popular. It's also very powerful. MySQL is the default database for a lot of common web projects and tools. The most popular blogging tool in the world, WordPress, almost always uses MySQL. Most developers are extremely familiar with MySQL and will often recommend it due to their comfort level and its ubiquity. It does come with a catch. MySQL is now owned by Oracle, so it has certain commercial use restrictions. If you have it installed on your server and your application talks to it all day long, MySQL won't charge you a penny. But if you want to bundle it up and sell your application so that someone else can install it on their system, you have to pay. It's still far cheaper than the previous options, but keep that in mind if you plan to sell an application that comes with a built-in database.

PostgreSQL

PostgreSQL is a completely free, completely open-source relational database. It's fast, feature rich, and (once again) free. This makes it my database of choice. The cons? It's not as ubiquitous as MySQL, and many developers are not as familiar with it. PostgreSQL is also a little tricky to install and manage, but there are services out there like DatabaseLabs (http://www.databaselabs.io) that offer cloud-based versions of PostgreSQL, so you don't have to worry about the setup.

You may have noticed that most of these databases include the letters SQL in their title. SQL (pronounced "sequel") stands for "structured query language." SQL is the language of most relational databases. It's how developers and database admins interact with the data inside of the database.

Time for a quick example. Let's say we had a database with a table of customers' information. If we wanted to list all the customers, we would use SQL to "ask" the database for the information like this.

```
SELECT * FROM Customers;
```

The asterisk means "everything," and "Customers" is the name of the table we're querying. That query would return something like this.

ID	Username	Email	Phone
12345678	bobsmith	b.smith@test.com	555-555-5555
12345679	supercoolguy	alex.c@test.com	555-555-5556
12345680	ms.direction	jane.g@test.com	555-555-5557

SQL is an old and robust language. It can do hundreds of extremely complex tasks. Most developers you'll meet will know SQL intimately and probably use it every day.

NoSQL Databases

Relational databases are still king in the development world, but there are a number of challengers to the technological throne. There is a broad and not very well-defined category of databases that falls under the heading "NoSQL." This is commonly interpreted as "Not Only SQL," because while they take a different approach to traditional relational databases, they often have query languages similar to SQL, and in some cases, they actually use SQL.

It's difficult to describe NoSQL databases because they cover such a wide range of variations. Some of them chunk data into "nodes," or "objects," or equally esoteric terms. This basically means that instead of column-and-row-style tables, they might have a block that holds all of one customer's information together.

For example, if you looked in one of the nodes you might see something like this.

```
{
    "name": "Bob Smith",
    "customerID": 12345678,
    "address":
    {
      "street": "1234 Lazy Lane",
      "city": "Tampa",
      "state": "FL",
      "postalCode": "33333"
    },
    "purchases":
    [
      {
        "orderNumber": 45678910,
        "product": "Big Green Thing",
        "orderTotal": 25.99
      },
      {
        "orderNumber": 45678911,
        "product": "Big Blue Thing",
        "orderTotal": 59.99
      }
    ]
}
```

If the developer then queried the database for Bob Smith, they would get all Bob's information back in one chunk. This is a very limited example, as there are so many different types of NoSQL databases. We would need an entire book just to survey them.

NoSQL databases carry some advantages and disadvantages when compared to relational databases. They are easy to use and extremely fast, especially for adding data. Their "chunk" nature makes them great for storing complicated data models, as you don't have to build Frankenstein's monster tables and relationships to model anything lacking a straightforward structure. On the other hand, they're still fairly new and can be a little rough around the edges in terms of available features, support, and expertise. There are also some technical issues with them, but I'll leave those to your developers to deal with!

There are tons of NoSQL databases out there. The most common are MongoDB, Redis, and Cassandra. If your development team recommends a NoSQL database, they might have a very good reason for it, but be sure to ask for an explanation.

Third-Party Integrations and APIs

Like everything else in this day and age, systems don't live in a vacuum. Your online e-commerce website most likely wasn't built from scratch, and your developers almost certainly didn't write the credit card processing functions themselves. This is because there are so many powerful services and software frameworks out there that half the battle is already won.

In 2001, a large national brand approached my partner to build a full e-commerce store. With Java as our language of choice, we set about the task at hand. We built a templating system that led each page on the website to have the same menus, logos, look, and feel, and made the site into one cohesive entity. We built a custom database to handle the specific products, descriptions, prices, weights, shipping costs, and a lot more. After finishing the database, we built the logic in Java to tie the frontend to the database, building shopping carts, user account management, and security systems from scratch. We then moved on to the credit card processing. We contacted the client's bank and worked with their developers on building an integration system between the site and their merchant gateway services. Finally, we helped the client through the credit card security and audit process. All in all, the project took eight months and cost $100,000 (it probably would have been a lot more if we had known what our competition charged—ah, the follies of youth).

As a counterexample, a few weeks ago, a client came to us wanting a custom e-commerce site. They had some complexities in their sales process that wouldn't work with the major off-the-shelf e-commerce systems abundant on the web. After working out their requirements, our programmers grabbed a development framework and the framework's most popular shopping cart module.

These were both free, open-source applications. After installing them, our designers tweaked the look and feel to make it fit in with the client's existing web branding. The programmers then went in and wrote a few custom modules to handle the intricacies of the unique sales process and installed the Stripe credit card processing service module. After the client activated their Stripe account, they launched the site and were selling the same day. The entire process took four weeks and cost under $10,000. The site was significantly slicker, more feature-rich, and probably more secure than the completely custom-built system from 16 years earlier.

Nerd Out

A **framework** is a set of software and database structures that have a lot of the most commonly used features a programmer could want. You don't want to reinvent the wheel every time you build a car, right? So why would you write a user login and registration system from scratch? It's already been done a million times. Frameworks have things like login systems built in, as well as many other useful features. Frameworks are often "opinionated," meaning they force the developer to code in the framework's style. As a result, some programmers fall in love with their frameworks, and others hate them.

The abundance of third-party systems and services like Stripe (and thousands of others) has freed up developers' time to build more advanced, interesting systems by relying on already well-established code. I don't think we'll ever write a credit card processing system again, let alone a shopping cart or even a templating system. These days, there are thousands of integrations available for you to piece together and leverage, allowing you to build the perfect, customized system for your business.

Want your software to integrate with your accounting package? There is most likely an API your developers can use to do that. Want to add a blog to your site? No problem. They'll install WordPress. Have your app's communications go through MailChimp? There's an API for that.

You get the idea. Just like in the Pure API Plays section, there are APIs for almost every service you could want your application to integrate with.

The API and You

In the same way that your system can talk to third-party systems through their APIs, many new systems include a built-in API from the start. This might sound strange, but I'll explain. Let's say you wanted to build an app on which users searched for available jobs in your company. We know there will be an administrative web system where your HR staff will add new jobs, along with their descriptions, pay ranges, and required experience. There will need to be a database to store all the job applications that flood in. On top of all this, there will need to be a mobile app to display the job data to the users and accept the applications.

Now, while we could build this all as one giant system, the modern approach is to build the whole thing like Legos. In other words, to build functional blocks that can all fit together to make something much larger and more complex.

Because we know that the mobile app needs to communicate with the administrative web backend, we can start by defining the rules they'll need to utilize to talk to each other. In our company, we do this by first defining an API, which is a set of web addresses that output very boring-looking data.

An example might be http://www.mycompany.com/api/list-jobs, which could conceivably return the following two, slightly odd, job postings:

```
[
  {
    id: "12345",
    title: "Sales Person",
    minimumPay: 30000,
    maximumPay: 50000,
    start_date: "ASAP",
    locations: [
      "Tampa, FL",
      "New York, NY"
    ],
    Description: "Looking for candidate with strong
sales experience in selling large monkeys to
upwardly-mobile NFL fans."
  },
  {
    id: "12346",
    title: "Sales Manager",
    minimumPay: 50000,
    maximumPay: 70000,
    start_date: "2016-01-01",
    locations: [
      "New York, NY"
    ],
    Description: "Looking for managers with experience
in primate sales. Hockey fans a plus."
  }
]
```

We now have a fairly human-readable output, regardless of how you feel about monkey sales. We design this bit first so the backend developer can start working on a system that outputs this API "endpoint." The frontend developer, working on the mobile app, can work on getting that data into her application and formatted to display it beautifully to the user.

When all is said and done, you might have hundreds of these so-called "API calls" that allow your frontend and backend developers to use two completely different languages to code their systems, but they have this very simple language in common to talk to each other. The other nice thing about developing this way is that when the project is finished, you could open up your API to the world. This permits other programmers to use your job listings in their own application—a win for everyone (except maybe for the monkeys).

Wrapping Up

That pretty much wraps up the most technical parts of this book. I hope it wasn't so overly complicated that you got bored, or so unbelievably basic that you felt patronized. This is a good lesson to learn: When explaining anything in software engineering, it's difficult to know the technical level of the person with whom you're speaking. My favorite line is, "I'm very nerdy, and I'm a programmer as well. How geeky are you?" That generally gets both parties on the same page, and their answer is a good indicator to me of how technical I can get.

PART II
START CODING

CHAPTER 5
WHAT CAN YOU BUILD?

"Why, sometimes I've believed as many as six impossible things before breakfast."
                        —Lewis Carroll, Alice's Adventures in Wonderland

This may seem like a strange question. I mean, imagination is your only limitation, right? Well, that's partly true. I often tell people, "I can make a brick fly if you have enough time and money," but what I actually mean by that is some things are easy, and others are expensive.

This section deals mainly with the types of applications you might consider and on which **platforms** to build them. A platform can mean a lot of different things to different developers. Most often, it's used to describe the kind of device an application runs on. The iPhone or Android operating system are platforms.

These days, there are the three major categories of "apps" you're likely to run into: web applications, mobile apps, and apps as consumer electronics. We'll review what these can do and some of the vocabulary you're likely to encounter when discussing them with tech folk.

The Software Market

"I have never worked a day in my life without selling. If I believe in something, I sell it, and I sell it hard." —Estée Lauder

When it comes time to build your application, you will need to consider what platform to launch it on. Platform choices include the web (and which browsers you're going to focus on), mobile devices (and which devices you're going to support), and a variety of secondary markets (wearables, smart TVs, kiosks, and so on).

The Mobile Market

Mobile devices come in many shapes, sizes, and technologies, but there are only four real players in the market: Apple (iOS), Google (Android), RIM (Blackberry), and Microsoft (Windows). These account for almost all the operating systems running on every mobile device out there. There are a few others, but most have such a small share of the market that they're not worth mentioning. At the time of this writing, the breakdown is around 17.9 percent for iOS, 81.7 percent for Android, and 0.3 percent for Windows (still trying to convince the world that it wants a Windows Phone). Clearly, Apple and Google have the market locked down at the moment. However, there is a little more to it than device market share.

The stats I like to look at are actually the number of apps available and the number of downloads, vs. revenue for each platform. Let's take a look:

The latest data from Statista's (http://www.statista.com) shows that the Google Play Store has around 3 million apps, while the Apple App Store has around 2 million. Pretty close, there.

App Annie's data shows that Google Play has almost twice the downloads from the App Store. This, however, does not take into account Apple's own apps (Pages, Numbers, Garage Band, etc.), which means the gap is probably a little smaller.

Here's the kicker: Apple users still seem more willing to spend money on apps and on in-app purchases. Apple generates almost 75 percent more revenue on apps in their App Store than Google.

This gap does appear to be closing fast, but it's still a big enough difference to consider when building a mobile app.

The Web

Device platforms make a much bigger difference in the mobile space than on the web. Browsers like Chrome and Firefox have more in common because they're trying to work to a standard.

According to the W3Counter (https://www.w3counter.com):

- Google's Chrome is first—currently the most popular browser on the planet with a massive 62.4 percent market share

- Apple's Safari browser comes second with almost 13.5 percent

- Microsoft's Internet Explorer (in all of its versions) is third with around 9 percent market share

- Firefox (my browser of choice) comes in at 7.8 percent

- The final few percentages belong to less mainstream browsers like Opera

Google's Chrome and Apple's Safari are normally the easiest browsers for which to design software. Firefox is also reasonably standards-based, so it is widely compatible with the more complex web tools out there.

Internet Explorer (or the new Microsoft Edge browser) is the red-headed stepchild of the web browser world. Microsoft has done its own thing for a long time when it comes to the web, although no one seems to be sure why. The various versions created a number of problems for developers. As a result, you might see an "Internet Explorer Tax" added on to a proposal from a development firm if you ask for IE compatibility. This "tax" is your developer's way of saying that they really don't want to deal with the hassle of making something work in IE. If you're working in the governmental space, you might not have the choice to ignore Microsoft's browser. You will probably have to pay extra.

Web Applications

What can you do with the web these days? That question might be better phrased as, "What *can't* you do with the web these days?"

The World Wide Web (its proper name, though no one calls it that anymore) is now so powerful and feature-packed that there are very few limits to what it can deliver. My personal prediction about the future of software development is that very shortly, every application you interact with will operate completely over the web. This goes for mobile apps, too.

Right now, native mobile apps are better than web experiences because they leverage the power of the device on which they run. The iPhone 6s and Samsung Galaxy S6 are mind-bogglingly powerful devices—more powerful than most computers were five years ago. I think in the near future, the "mobile web" will become the main platform of choice to deliver what we now think of as apps. This transition is now complete in the desktop market, where web versions have replaced most traditional software.

Web Capabilities

The term **web application** is the de facto language used to distinguish systems with a high degree of complexity and interactivity from the more mundane "website" label. Basically, they're the same thing: A website is a simple web application, but it's also a catch-all phrase. The best way to think about it is that one defines a website by its content (it is an information delivery system, like a brochure in cyberspace) whereas one defines a web application by its functionality (literally, its application—the use you get out of it).

Websites typically have pages of information linked together, allowing the user to navigate to other pages that contain other bits of information. Web applications are often a single page, but that page has a lot of interactive features on it. The line is frequently blurred, but that's a good enough definition for now. Terminology aside, web applications usually have a few distinguishing features:

1. They are dynamic, meaning you can add data to the system, save it, and then manipulate it in some fashion down the line. This means that a web application has a database sitting in the background, storing the user's inputs, enabling editing and saving.

2. User login systems. Probably the most common feature across web applications is the ability to register somewhere in the system with your email address and a password so that you can come back and access content, data, or features specific to you. Once again, this means storing user information in a database.

3. A specific utility. The application has a purpose other than just displaying universally accessible information. A web application generates the information you see for your specific case. It might be as simple as pulling your personal details out of the database and populating a form for you to print out. Or it could mean that the system allows you to do something as complicated as online graphic editing. Pixlr (https://pixlr.com/) is a web tool that is almost as powerful as the ubiquitous Adobe Photoshop, and it's completely free.

"Rich" Web Experiences

Originally used to describe Macromedia (later Adobe) Flash applications, **rich web experiences** are now a diverse group of technologies and web applications that offer a desktop-like experience to the user. The web is entering its middle age (as far as a technology goes) and as a result, it's able to handle a lot. Browsers and hardware, too, have come a long way, allowing for delivery of everything from photo editing software to interactive 3D games over the web. A lot of this is due to the increasing power of JavaScript. The other major factor is the power of browser extensions and plugins. These enable the humble (yet now very powerful) web browser to learn new tricks all the time. For example, the Unity game engine, which allows developers to build 2D and 3D games for everything from iPhones to PlayStation, has a browser plugin that lets gamers play the latest 3D graphics games right in their web browser.

Desktop applications are not dead by a long stretch, but given the development, distribution, and maintenance of rich, web-based applications, things do not bode well for traditional applications.

The Mobile Web

The web is everywhere now, but most importantly, it's on phones. In 2014, mobile overtook the desktop in Internet traffic. That means more people access the web via their smartphones than on PCs, Macs, and everything else combined. The problem is that most websites and web applications were never designed for small screens. Having to zoom and scroll around a web page on mobile is extremely frustrating. There are two solutions to this problem. Each involves optimizing the experience for mobile users.

A dedicated mobile site

While not as popular as it once was, you can build a web experience purely for the mobile web. This is a web application designed purely to work well on a smaller, portrait (rather than landscape) screen. Usually, this is done with the existing desktop version unchanged. In other words, if you go to http://www.some-website.com, it is designed only for large screens. However, if you visit that same website on a phone, it will detect that you're on a mobile web browser and redirect you to something like http://m.some-website.com (the "m" being the standard mobile indicator). This site, if viewed on a desktop screen, would look pretty weird—stripped down and strangely stretched out. On a phone screen, it looks great!

The benefit of a dedicated mobile web application is that the developer has a lot of design control over the site. They can optimize the views and pages to a very large degree of detail and present the user with an app-like experience over the web.

The downside, of course, is cost. Essentially, in this model, you're building two web applications. The cost might not be as high for the mobile version if there is already a desktop version (the mobile version is using the same data, after all), but it's still a complete redesign.

Responsive design

The buzz phrase of the last few years has been "responsive mobile design." This refers to websites and applications that look natural on both a desktop and a phone screen. A great example of this is The Boston Globe's website (http://www.bostonglobe.com/). You can play around with responsive design by visiting a website, then dragging the window of your web browser to make it tall and skinny rather than widescreen. This emulates the various changes a site goes through as users view it on progressively smaller screens. In order to achieve this effect, the developer has to program specific break points for different screen sizes. Once reached, the developer switches out their layout code. Combine this with some automatic image scaling techniques, and you get a web page that transitions smoothly between a full-sized web application into a mobile app simply by resizing the window.

Though not as difficult as it used to be, executing quality mobile responsiveness can be quite technically challenging. The specific sizes of the devices you're targeting radically alter the time and cost of development. Most companies are happy to have a system that looks pretty good across all screens. In those situations, you have a few break points (usually desktop-size, tablet/iPad size, and average mobile phone size). If it doesn't look 100 percent perfect on every device on the planet, it's a negligible gap. However, we have clients who require ten or more break points. They may even hire a design firm to make mockups of each and every screen size and show what the expected look and feel is on each and every device. This requires a huge amount of time and energy. It's normally only tackled by enterprise-sized companies, and only the more technologically sophisticated ones at that.

Mobile-First

Due to the huge increase in mobile phone usage for apps, web browsing, and gaming, many companies and industry thinkers push the "mobile-first strategy." Instead of first designing and developing a desktop experience, then altering it work for mobile, we start the other way around. This might not strike you as a groundbreaking idea, but remember that everything designed and programmed for mobile devices is still built on traditional desktop or laptop systems. It sometimes seems more natural to design for the same platform you're designing on.

It makes a great deal of sense to design for mobile first, mainly because it's easier. If you can make something look good on a small screen, it seems to be more intuitive to scale it up in size to fill the space on a larger screen. Thumbnail-sized images grow into banners, and horizontal deciding lines on a phone screen expand into full boxes of content when viewed on a TV. Mobile-first should be the way you go, as well. When your developer starts wireframing your application, be sure that the first screens you see look good on a phone.

Mobile Apps

Web applications are quickly replacing traditional software in almost every realm. The improvements in web technologies, specifically the amazing strides in browsers and JavaScript allow developers to build everything from accounting systems to games in the cloud. Still, traditional software is not entirely dead. In fact, it has found new life in the mobile market space. Apps are everywhere. Most of the time, they are frontends for software delivered over the Internet anyway, but many of them utilize the massive amounts of power of modern phones and tablets to do amazing things.

Examples of frontends of website apps you might be most familiar with are the LinkedIn or Facebook apps. They are full-featured mobile apps, built with traditional software development techniques, but they don't really do many things on their own. They are fast, good-looking frontends for content fed to them via the web. What you see in the app is almost exactly the same thing you would see on your web browser. They might do a few extra things (like upload photos directly from the camera on your phone), but, in essence, they are a window on your phone that looks at the website over the Internet.

Mobile games, multimedia creation and editing software, and home automation systems are typically "web enabled," meaning they use the Internet for certain communication features, but most of the power exists inside the app itself. These are typically much more complicated to build and leverage all that power on a modern device. 85 percent of all software will come from mobile apps by 2019, so if you don't have a mobile strategy of some sort, you need to start thinking about it now (unless your target market consists solely of rural, eighty-five-year-old Amish wood carvers).

What Can an App Do?

Mobile apps can do quite a lot these days. Everything from full-blown 3D games to augmented reality apps are possible with the powerful hardware available in your pocket. Developers leverage a number of tools built into these devices. For example, an app could use the longitude and latitude coordinates from the phone's GPS system to determine if the user is within 100 meters of their worksite. If they are within that radius, the app would allow them to punch in on their time clock.

A smartphone also has a lot of practical software features running on it, built by Apple, Google, Microsoft, or Blackberry (depending on your device). Developers tap into this software to improve applications. Examples include Apple's "in-app purchasing" system and their TouchID fingerprint authentication system.

There are limitations to what an app can do, dictated by the provider of the phone or tablet. Most developers can't access lower-level features of the device. I get asked all the time if I can build an app for a client that disables text messaging if the phone is moving faster than a few miles an hour. The intention is that a parent could install the app on a teenager's phone, and they wouldn't be able to text-and-drive. The problem is, I can't turn off text messaging. Ever. I'm sure if I was AT&T or Verizon, Apple would give me access to those features (for a price), but for everyone else, they are off limits.

The following is a list of some of the standard features available on most smartphones and tablets. This should help kick-start your creativity if you're trying to come up with novel uses for an app that leverages the technology built into the device.

Camera and Add-ons

Apps can access the camera on smartphones and tablets, as well as the library of photos on the device's image gallery. If the phone has a set of image effects (like sepia tone, inverted, black and white, etc.) the app can also use those effects while in camera mode. Images can be taken and then brought into the app to manipulate or save.

Accelerometer

Most mobile devices have an accelerometer built into them. An accelerometer is a device that measures, you guessed it, acceleration. This allows the device to know if it is moving and at what speed and orientation. A coder can use the measurements from the accelerometer to determine whether a phone is in landscape or portrait mode, if a phone is being shaken, or if the phone is being tapped vigorously. There are even apps out there that let the user "paint" in midair by waving their phone around. They use accelerometer readings to draw beautiful, glowing lines on the screen as if by magic.

Geolocation

We recently built an app for a vintage Volkswagen Beetle specialty shop. They wanted something to hand out at a trade show that stood out a little more than a novelty pen. Our solution? An app designed to look and function exactly like the speedometer on a classic VW. As you can imagine, a lot of things can start going wrong with physical speedometers over the years, so why not replace one with your smartphone? The app had the repair shop's phone number and a map to their shop built in, as well as the ability to track your mileage. It was a massive hit and was surprisingly accurate. It worked by looking at the GPS coordinates of the phone and calculating the speed based on distance traveled.

Most of us are familiar with the GPS system on our phone. Like the microwave oven, Google Maps has become one of the few technologies I struggle to imagine life without. Generally, a developer can use a phone's GPS system to access the location of the device and use it for their needs. This may involve displaying the user's location on a map, tracking their movements, calculating their speed, or "geofencing" an advertisement. Geofencing is a fancy term for drawing a perimeter around a point and telling the device to do something if the user goes into that circle. A good example is the time-tracking tool discussed earlier for hourly employees. These employees punch in once they are in a certain geofence—so they can't work from home or punch-in while watching Netflix on the couch.

TouchID

Fingerprint readers now enjoy increased popularity on laptops and phones. They are handy for logging a user into a system without needing to type their username and password or for adding an extra layer of security.

Contacts

With a user's permission, an app can access a device's contacts list or address book. Imagine an application designed for trade shows. You scan the code on a participant's badge, and their contact information, as well as a note explaining where you met them, is automatically imported into your address book. Imagine an application for a social media site that automatically connects you with people you already know by searching the email addresses of contacts in your address book.

Headset Detection

Headset detection is one of the less commonly used tools on modern devices, but I've always thought that it has a number of interesting utilities. A few years ago, I participated in a startup competition where one of the most interesting apps presented was a security app targeting college-aged women. The app triggered a text message to the user's emergency contact, automatically dialed a security center, and relayed the user's GPS coordinates in the event that the user yanked the headset out of the audio jack. The idea was that if a user were out jogging and felt they were in danger, they could yank their headphones out of their phone, which would trigger the app's security systems.

In-App Payments

In-app payments allow developers to charge users for stuff inside an application without asking for a credit card number. The app asks for permission to use the credit card already on file with the app store. This way, the developers don't need to set up complicated billing systems. They use ones already provided by Apple, Google, or Microsoft. The only hitch is that they'll take 30 percent of whatever you sell!

Whether you're paying $3 to remove all those annoying ads in an app or buying whatever "lollypop hammers" are in Candy Crush Saga, in-app payments are the way it's done. In-app purchases allow users to buy virtual goods for games, unlock new levels, remove advertising, or upgrade to premium features. There are a lot of rules and regulations for each app store regarding the use of in-app purchases, but there are two simple rules of thumb:

1. You can't charge for real-life items or services with in-app purchases. If you buy something from Amazon's app or use Uber, you'll have to enter your credit card directly into the app.

2. You can't handicap the application in some way, like only granting access to certain features (for instance, allowing a user to save one journal entry a day in a diary app and then charging them if they wanted to add more than one). For some reason, this rule does not apply to games. You can charge to unlock new levels, weapons, vehicles, or pretty much anything you can dream up.

The rules are fairly complicated, and you'll want to discuss what exactly it is that your in-app purchase will do with your developer and possibly an attorney.

Calendar

Every smartphone or tablet comes with a convenient calendar application. I live and die by my calendar app, so it's extremely useful that I can add and share events from it as well. Like the contacts list or address book on a phone, the calendar is accessible for apps, with the user's consent, to make changes to their schedule. A recent medication tracker we built for a national cancer research hospital automatically adds the patient's appointment reminders to their calendars. The app also sets alarms to remind the users to take their medication and does so intelligently by checking to see if there are any conflicting appointments already in their calendar. This way, you don't have a medication reminder going off while in a meeting.

Scanning Barcodes and QR Codes

Barcodes and QR codes (Quick Response Codes) are pretty much everywhere these days. Barcodes are most commonly seen as Universal Product Codes (UPCs), the 12-digit product ID numbers you see on items at a store. QR codes are two-dimensional barcodes, meaning they have vertical and horizontal lines. This allows them to store things like URLs of websites, email addresses, and even contact cards, complete with phone numbers and addresses.

Most devices these days include built-in systems that allow developers to read barcodes or QR codes. If they are not standard to the device, they can easily be integrated into an app.

QR codes store a lot more data than 12 digits, depending on their density and how you encode them. Theoretically, you could encode two pages of this book in a single QR code, but it looks pretty intimidating:

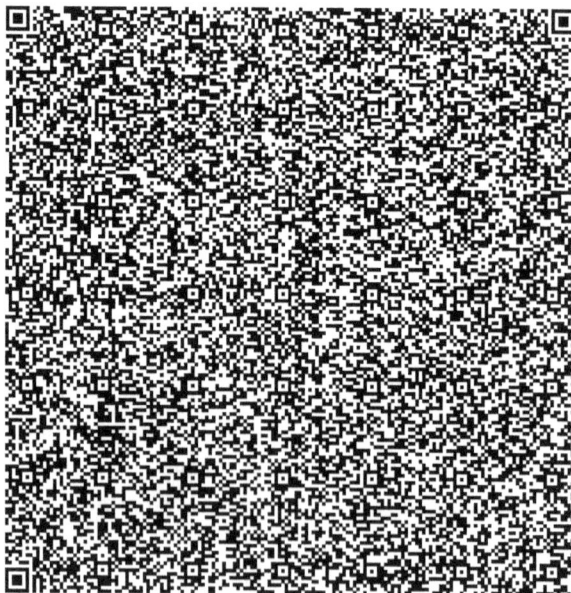

Push Notifications

These are the little text popups that appear on your device like text messages but disappear as soon as you acknowledge them. My favorite push-notification-enabled application is the American Express app. Every time I make a purchase on my Amex card, I instantly get a push notification telling me the location and amount of the transaction. If I don't recognize the merchant's name, I tap the notification, and it opens the app for me to investigate further.

You can use push notifications for just about anything. Unlike text messages, they cost nothing to send, and they're less invasive than email. The downside is that they can be overused and quickly become annoying. I know I haven't gone to the gym today, but you can stop reminding me, already . . . I feel bad enough as it is!

In-App Browsers

A common requirement for an app is the ability to link to a piece of information on the web. Let's imagine an app that allows you to plan commuter train journeys. You purchase tickets in the application with a credit card, and your schedule is nicely synced with your calendar. However, each train station might have its own website with essential information not available in the app. In this case, the user clicks on the "Web" icon, which takes them straight to the website. This happens in one of two ways:

1. Clicking the "Web icon" opens the device's native browser (Safari for iOS and the Android Browser or Chrome for Android devices) and loads the content. Our train app slips into the background, and the native browser pops open.

2. We use an in-app browser to open the website. This is the same underlying technology as the native browser, but it is completely controlled by the developers of the app itself. Using this technique, you would literally have to program a close button or a toolbar of some sort. Otherwise, there would be no way for the user to get from the browser back into the app.

This technique is particularly useful if you have a section of your app frequently updated with simple content. For example, if you had a web forum for

enthusiastic train travelers, you could have one of the menu items in your app
be "Forum," and tapping on it would load the "web view." You can have a web
browser embedded into any part of an app by designing the website to match the
app. The user wouldn't even know that half of their screen showed a website and
the other half, the app's menu, header, and so on.

Phone Dialer

Yes! Your smartphone also makes phone calls! Apps, too, can trigger phone
calls or bring up the phone's dialer. If your business has an app that shows you
the closest location to a sales office, it would be useful to click on the location's
phone number and have the phone call that number. Apps also generate text
messages. A common request from our clients is a "share" feature, where you can
send prewritten text messages to your friends telling them how cool this app is,
along with a link to download it.

Proximity Beacon

Low Energy Bluetooth (BLE) is a technology that is starting to appear in the
wild more and more often. Proximity beacons enable your phone to read small,
short-distance radio signals, beeping out a single code into the world. Your
phone determines your distance from these signals based on their signal strength,
and they can even triangulate its relative position between a number of these
"beacons."

Beacons are small, plastic boxes varying in size from a matchbox to a hockey
puck. They can be battery powered, run off a wall socket, or USB powered. A
beacon is not a very smart device. It simply sends out the same signal again and
again. Each code is unique to that beacon. It's up to an app on your phone to
recognize this code and then do something in return.

Let's build an imaginary app for an imaginary chocolate shop. We will buy two
beacons, a blue one and a white one. The blue one puts out the code A000123,
and the white one B000456. That's all they do, day after day, year after year, just
yelling their code out into the abyss. We'll stick the blue one on the door of our
shop and the white one near our specials display (even though all chocolate is
special). Our code will specify that if it hears the A000123 (the blue beacon), it
will push a greeting to the user as they walk into the door. "Hello, Jane! Welcome

back to Chocoholics. Be sure to grab a free sample of our new French nougat and walnut chocolate bars." We said "Welcome back" because we registered Jane's first visit a few weeks earlier.

If Jane opens the app, it shows her a map of the shop, with a blinking light over the specials table and a "Nougat Finder Bar" at the bottom of the screen. The louder the signal of "B000456" from the white beacon, the closer Jane's phone is to the specials table. The bar could fill with walnuts and chocolate the nearer to the specials table Jane walked. On her way out, the app could detect that Jane left the store, and it would generate detailed statistics about the length of her visit, how much she spent, and what constitutes her favorite chocolates.

Okay, that was a pretty detailed scope for something that will never exist and is probably overkill for a small chocolate shop. But chocolate inspires fantasy, so you can't blame me.

NFC

Near Field Communication (NFC) is another tagging system similar to barcodes, QR codes, and proximity beacons. NFC is similar to proximity beacons in that it works off humble radio signals. The "tags" are a lot like the RFID tags you'll find on theft protection systems in clothing stores. They are basically stickers with some concentric metallic bands printed on them. Those metallic bands are actually the stickers' antennas!

Unlike proximity beacons, which use Bluetooth, NFC only functions if your device is very close to the tag—usually ten centimeters or less. Think of it as having to touch the sticker with your phone to get it to work. Touching the tag can then trigger a function on the device, like putting your phone into sleep mode when it touches the NFC tag on your bedside table, or launching Google Maps and showing a pin of your restaurant's location when a customer taps a poster advertising the newest location.

NFC is a common feature in most modern smartphones, but it isn't available to developers on Apple-based systems yet. Apple uses NFC for their credit-card-like Apple Pay, but beyond that, they have it locked down for now. As a result, the NFC apps we've worked on were Android-only and were internal business applications like inventory control, package tracking, and timesheet punch-in systems. Apple may start to open up NFC to developers, opening up a very large market, but this is pure speculation at the moment.

Native vs. Hybrid Applications

Now that you have decided what you're going to build and what your application is going to do, you'll want some idea of what kind of developer you'll need.

There are three main options available to you when developing a mobile app: Native, cross-platform, and hybrid. So, what's the difference?

If you do a quick Google search for "native versus cross-platform versus hybrid apps," you will end up with more opinions than could possibly be helpful. In this section, I'll summarize each option and discuss their pros and cons, as well as some of the technologies involved. Feel free to explore the various technologies at your . . . leisure?

Native Apps

A native application is one that doesn't need any external systems or frameworks to run on your device. It is written in a "lower level" programming language, meaning that it takes less work for the device's processor to understand what to do with the code. In almost all circumstances, native applications run faster than their rivals. Native languages are what operating system companies like Apple and Google write their systems to work with, giving developers direct access to things like 3D graphical acceleration and built-in user interface components.

Native technologies include:

- Java (the native language for Android)
- Objective-C (the native language for iOS)
- Swift (Apple's newest native language for iOS)
- C++ (the native language for Blackberry . . . but it's more complicated than that)
- C# (the native language for Windows phones)

Pros of native apps:

- Nothing is ever going to be faster than a well-written, native app, period.

- Native application developers get first access to the newest bells and whistles from Apple, Android, Microsoft, etc.

- Native applications have a more natural feel. This is subjective, but it goes back to the speed issue. If your users access your app several times a day, a minor delay in a tap or a swipe might grate on them over time.

- There might be a feature on a device you want to use that is not supported by anything other than the native language, which means a native app is your only option. I have run into this problem once or twice in the last few years but only for very specific applications.

Cons:

- If you release your application on more than one platform, which is often the case, you may need more than one development team. Software engineers, like everyone, tend to specialize. It's rare to find a high-quality Java programmer who is equally at home using Swift. If you're using an agency with both developer specialties under one roof, issues regarding resource scheduling between the developers might arise.

- It can often take longer to design something natively. There are basic layout patterns for every platform, but doing something innovative or slick can take a lot of time to perfect. It's tough for a native developer to immediately see the result of their work. They have to compile their code, which can take a few minutes, and then debug layout changes in the compiled code. (This step gets very tricky.)

- Time, complexity, and specialization mean that it will almost certainly be significantly more expensive to go the native route. If money were no object, you could hire developers for every system your application could possibly need to run on. Unfortunately, I've never heard the line is no object" outside of the movies.

Cross-Platform Applications

Cross-platform means that a piece of software can run on two or more platforms, which could refer to hardware platforms or software platforms. This is not a great definition, and as a result, people in the development world tend to throw this term around quite a lot, myself included. What a coder means when they discuss a "cross-platform" application is a website and a web browser in one. Wait. . . . What?

HTML, JavaScript, and CSS applications (or websites), as we discussed in The Anatomy of an Application: Frontend Development, can run on pretty much anything that has a web browser. So, why not include the web browser and install the whole thing wherever you want? All you would need to do is make the "website" look more like a traditional app (easy), then get a special web browser for each platform.

Sure enough, this model does exist in a number of varieties. There are several ways of doing this, but they mainly boil down to the website and web-browser combination system. These browsers, called "headless browsers," don't have things like address bars, "Back" and "Forward" buttons, or bookmarks. In effect, you don't even know that what you're looking at is a web browser—because it looks like an app. Clever software can bundle these together, compiling the two so that you can upload the cross-platform app to the app store of your choice or deploy it on custom hardware, including everything from watches to airline entertainment systems. A ton of APIs exists for your application to interact with all the device-specific features, like GPS systems, cameras, accelerometers, etc. A cross-platform app can do pretty much everything a native application can.

Cross-platform applications are extremely common these days. This is mainly due to the fact that web technologies became significantly easier to learn than native languages over the course of the last two decades. There are a lot of web developers out there who made the switch to the cross-platform market with ease.

As is the case when anything is easy and common, however, there is a lot of garbage out there. As the web and cross-platform systems evolve, things become increasingly complex. Bad code written by inexperienced or untrained programmers is common in this space. For this reason, it is often looked down upon by native developers. Native coders are more likely to have come to programming through more traditional paths, like computer science degrees. Cross-platform developers more commonly from the web development community, which is significantly more varied. It's no surprise that these two worlds clash.

Don't get me wrong, there are great developers in the cross-platform space, both from traditional computer science/engineering and web development backgrounds. They take what they do very seriously and write exceptionally good code. Netflix develops their interfaces in this fashion and, needless to say, they know what they're doing.

Cross-platform technologies include:

- HTML 5/CSS 3/JavaScript
- Ionic (http://www.ionic.io)
- Sencha (https://www.sencha.com)
- jQuery Mobile (https://jquerymobile.com)
- Onsen UI (https://onsen.io)

Pros of cross-platform applications:

- They always require a smaller development team and, as a result, cost less.
- They are iterated more quickly because they use much simpler layout tools, so turning designs into reality is significantly simpler.
- Maintenance is often easier, with changes rolled out quickly across devices.
- You can run them on anything, from a smart TV to an iPad to a full desktop application.

Cons:

- They are slower than their native counterparts because the app is actually rendering a web browser, then running all the code inside of that web browser. In addition, every time you make an API request to the native functions, the JavaScript request requires translation into the native language. The response is then translated back to something JavaScript understands.

- When Apple or Android (or any of the other big players) make changes to their systems, they don't consider cross-platform apps as much as native ones. Occasionally, this leads to operating system updates that can break your app. Of course, this happens with native apps as well, but not as frequently.

- When operating system updates come about, native apps get the newest features and design changes first. The cross-platform community must develop their tools on top of the native tools, so they tend to lag a few months behind native apps in terms of the newest bells and whistles.

Hybrid Apps

Many consider the terms "hybrid app" and "cross-platform app" interchangeable. They are, however, used quite distinctly by coders, bloggers, and pretty much anyone with an opinion. I'll throw my two cents in here as well.

The important thing to keep in mind with hybrid applications is their intention. They're trying to combine the best of the native and cross-platform worlds—hence, "hybrid"! That wasn't so hard, was it?

Anyway, here is my definition: Hybrid apps are a sub-type of cross-platform application. They attempt to do the following:

1. Use technologies not native to a device, and compile them to native code. By this, I mean you could get a native application (the equivalent of something written in Objective-C for iOS or Java for Android), by writing in a language other than Objective-C or Java. It sounds strange, I know, but these systems use some very clever software to do something loosely similar to translating a spoken language. In this way, they are distinct from most cross-platform technologies in that they don't require a "web view" or that built-in web browser we discussed earlier.

2. If they do make use of the web view model, they can blend it with native components. For example, the menu and navigation systems of an app might be native, so they can work as quickly as possible, but the pages that load from those navigation systems might use a web view.

Their intended goal is to let a developer, who is familiar with one set of technologies, publish their work to more than one platform while maintaining as many of the benefits of a native app as possible. This does not necessarily mean they are truly cross-platform—it might be the case that a developer only has to know one set of tools but still has to write two or more separate applications. These two applications would share a lot of code but would share less than the web browser-bundling method.

That might be a long and nuanced definition, but it covers the bases. The takeaway is that a programmer skilled in a hybrid programming system can theoretically build an app as complicated and speedy as any native programmer—with the added benefit of deploying the app to more than one platform. At the very least, you won't need multiple engineering for different platforms.

Hybrid application technologies include:

- QT, or "Cute" (http://www.qt.io)
- Facebook's React Native (http://facebook.github.io/react-native)
- Appcelerator's Titanium (http://www.appcelerator.org)
- Microsoft's Xamarin (http://www.xamarin.com)

Pros of cross-platform applications:

- Hybrid apps have many of the benefits of native applications, such as native user interface components. This means your users will feel more at home in a hybrid app than in a purely cross-platform version.
- You have most of the benefits of native application speeds.
- A single engineering team can be familiar with one technology and produce various platform versions of a particular application. This drastically cuts down on team size requirements, costs, and learning curves.

Cons:

- Hybrid application frameworks are significantly more complicated than good old-fashioned cross-platform apps.

- Developers familiar with them may be less widely available than cross-platform or native developers.

- A hybrid app system might only deploy to one or two platforms. They target iOS and Android, but not much else (yet).

Conclusions

With a lot of options, and even more opinions, it can be hard to work out what technology is right for you. Asking a programmer is not always going to be the best option, either. Native developers can be understandably elitist—they have one specially crafted tool to do one thing exceptionally well. To the native developer, a cross-platform application might seem like heresy. On the cross-platform side, developers can be overly defensive, claiming that their set of tools will solve every problem out there. The truth is somewhere in between.

Here are a few recommendations for making your decision:

1. HTML5/Cross-platform apps will never perform as well as native applications. In 80 percent of cases, this won't matter. It does matter if:

   - You have a seriously complicated user interface

   - You have high-performance requirements for your application

   - You want the absolute slickest interface possible

2. Any app can be of high or poor quality. Yes, you will save some time and money on cross-platform applications or hybrid apps, but that shouldn't be the main reason to use them. Go with cross-platform if you need to iterate quickly, or if you need to run your application on a wide variety of systems. Go with hybrid if you want to build native-style applications with a smaller core engineering team. Hybrid is my current personal favorite paradigm.

3. Remember that most applications, mobile or otherwise, are only
    50 percent or less user-facing. A lot of the work in any modern
    application is done on the server side—the app just sends requests
    back and forth from the server and shows the user a pretty interface.
    These are the kinds of applications most common in business cases, so
    the technology chosen to build the interface doesn't matter as much,
    and you can replace it down the line. If the app does more of the work
    on the device itself (for example, a photo editing application or a 3D
    game), then you'll probably want to go native.

Apps As Consumer Electronics

When we think of consumer electronics, we think of digital meat thermometers, GPS navigation boxes, and heart rate monitors. We think of a plastic box with a few electronics inside it that does one or two simple jobs. The kind of software running on these devices is low-level and complicated, but that's all rapidly changing. These days, the software that you write for your web browser might soon power your programmable coffee maker!

Apps for Everything Else

Most of us are familiar with the type of **apps** we use on mobile devices. The only reason we call them "apps" is because it's the diminutive form of "applications." There really is no difference between the creation of apps and the creation of any other type of software. For the sake of simplicity, let's define "apps" as software that runs on Android, iOS, and some of the other mobile operating systems out there.

But software is everywhere, right? The gas pump you use once a week runs some sort of hidden application, as do cash registers, digital signage systems, car stereos, lighting networks, and a million other hidden systems. Traditionally, this is the realm of "embedded software." Embedded systems are purpose-built computers, usually pretty small, designed with only one function in mind. These computers are programmed using very complicated programming languages because they're not very powerful. The more "native" the code, the quicker they run. This made a lot of sense when building these devices was expensive and you had to squeeze every last bit of power out of them to remain cost-effective.

Times change. Today, there are thousands of small factor computers out there, the size of USB memory sticks, that have a ton of power. Android is now the operating system of choice for many of these manufacturers. It's freely available and runs on lots of different types of these computers. This opened a floodgate of developers who can now build very interesting applications—the type of applications that, up until recently, belonged in the domain of the software uber-engineers.

Recently, I built an application for a company that has remote monitoring systems for animals entering and exiting barnyard entrances. Five years ago, the company would have had to invest hundreds of thousands of dollars and at least two years in research and development to design a computer control system that could hook up to their pressure plate sensors, waterproofing, and hardening systems. They would struggle with the device reporting back to their analytics systems.

This can now be replaced with a $30 Android stick computer with built-in WiFi stuffed into a waterproof box we bought from Amazon. The pressure plate connects via a USB cable, and the reporting system is just a web application we wrote in three weeks. All in all, from design to implementation, the first live market test took three months. The entire hardware system cost under $60 per unit. The Android system runs a small app that users can update remotely as new features become available.

We refer to these types of systems as "headless" because they don't have a graphical output. There is no screen or display hooked up for the cows to see what is going on—the app just reports back to a web system with which the farmers interact.

On the flip side of these tiny embedded devices are the full-blown computers needed to run more complicated systems. A good example of these are the interactive kiosks you may use at the mall or a digital signage system at your local sports pub. These brightly colored, highly visual systems generally have full PCs or Mac Minis hidden behind them running the show. The software running on top of them may be something more like the apps we've covered so far, but normally, they are traditional software systems. The advent of small form factor computers, running operating systems like Android, is radically changing this space as well. A Mac Mini running an interactive kiosk on a cruise ship costs $500 per box. An Android stick computer that can do 90 percent of the things the Mac can do might cost $50. A $14 version easily replaces a $400 PC running a digital signage system.

Software development and maintenance costs have also plummeted. What was once a highly specialized field is now full of common, easy-to-code technologies. We recently developed an interactive kiosk system for a trade show client in three days! The kiosk is just a website that runs locally on the touch screen and not on the web. Developing this way enables us to leverage commoditized technology (the web) to replace the once complicated traditional systems of only a few years ago.

The approach of using the web and mobile technology combined with cheap, readily available hardware to replace complex and expensive systems is currently a wide-open market for innovation.

Your Internal and External Systems

"When the number of factors coming into play in a phenomenological complex is too large scientific method in most cases fails. One need only think of the weather, in which case the prediction even for a few days ahead is impossible." —Albert Einstein

Many projects rely on more than just the little part on which you personally work. A sales team application may rely heavily on your CRM. An online ordering system needs to integrate with your point of sale, and an invoice and billing system needs the ability to export to your accounting package. Because of all these added complexities, you should spend some time on an internal systems audit.

This isn't some sort of IRS audit, but rather a simple document that you'll want to be able to hand over to your developers to make their lives a little easier and to give you a better overview of your new system, end-to-end. The following is our company's internal/external systems checklist, shared here for you to use for your own purposes. You can download a copy of this checklist by going to https://www.sourcetoad.com/herding-cats.

Internal/External Systems Checklist

General Documentation

Documentation is vital for a developer interacting with a third-party system. A system that isn't well-documented is pretty close to useless if you're trying to integrate it into a custom application. This is because there is an infinite amount of possible commands that one system may have, and there is simply no way a programmer can guess what they are.

URL to documentation: _____

Integration Method

There are three common methods of integrating data from one application to another. Find out which, if any, your existing internal and external systems use.

An API?

The best and often easiest method of integrating two systems is with an API. If you recall, an API is a set of commands that passes between two systems, typically over the web.

Is there an API? Yes No

URL to documentation: _____

Communications protocols?

If you try to integrate with a system that does not have an API, it might have another way to communicate that doesn't involve making easy web commands. This takes the form of anything from a piece of hardware that streams weather data in a raw format, to a forty-year-old network protocol.

Any protocol information available:

_____

_____

_____

Scraper?

Drastic situations call for drastic measures. If there is not a well-formatted, well-documented data transfer system on a system you need to integrate with, you might have to scrape the data. This means you'll need to develop a custom piece of software to grab the data from its source (usually a website) and intelligently format it so that it's usable.

Example URLs to scrape:

_____

_____

_____

One of the problems with scrapers is that your software must act like a human being reading the data. If you're visiting a particular website every five seconds to get new data, the website owner is probably going to ban you for overloading their server. You need to make sure the data is either freely available for reuse or that you have the rights or permissions to use it.

Describe any scraping risks:

_____

_____

_____

_____

Support?

Most of the time, you will have a support agreement in place with either the external vendor, who provides you with a system, or an internal team, who will play the support role for your new developers. It's really helpful to have all this information on hand, right at the beginning.

Support contact details:

Company: _____

Contact Person: _____

Email Address: _____

Web Address: _____

Phone Number: _____

External companies may have various support requirements that will affect your developers' ability to get that support. This might include certain times of day when support is unavailable or only five hours a month of telephone support. It's important to find out what level of support vendors and internal teams are willing to give before handing over their phone numbers to outside development teams!

What are the support terms?

_____

_____

_____

Contact information for third-party vendors?

In the same way that a product or service you use might have a dedicated support system, custom systems you've developed in the past may have some level of support from those previous developers. In an ideal world, you still employ the old set of programmers, and the new team will work on something else. This way, there is zero friction between the teams arguing over territory. Inform the teams who built any existing software that you are utilizing their system to build something new. Make sure they're ready to help support the new teams.

Support contact details:

Company: _____

Contact Person: _____

Email Address: _____

Web Address: _____

Phone Number: _____

What, if any, are the license restrictions?

Depending on your agreement with previous vendors, you will want to let your developers know what license restrictions they may face. These might include restrictions on how many API calls you may make in a day, how much data you can download at any given time, or even how you can use the data. A common example is showing the logo of the data vendor whenever someone uses the data.

List any concerns with the license agreement.

_____

_____

_____

_____

_____

This should help you get off to a good start and prepare you for what is possible and what is not. If the data to do what you want is unreachable, then prepare yourself for an uphill battle. If there are well-documented APIs out there, you can browse through them and get new ideas about what your software will be able to do.

CHAPTER 6
FINDING DEVELOPERS

"That's the thing about people who think they hate computers. What they really hate is lousy programmers." —Larry Niven

At this point, you know a little about the technologies involved in creating your application and have a pretty good idea of what you want to build. The next step is to find people to bring your plans for global domination to life.

Developers come in all shapes, sizes, skill levels, expertise, and price. In this section, we'll look at a number of options open to you for hiring these programmers. We'll also discuss a few tips and tricks to utilize along the way.

We'll primarily focus on how to do the following:

- Hire an internal team
- Hire an agency or development studio
- Hire a freelancer
- Hire an offshore development team

Let's get started!

Hiring Developers

"The city's central computer told you? R2D2, you know better than to trust a strange computer!" —C-3PO

If you represent a company, your first instinct might be to hire one or more full-time software engineers. In tech companies, the "engineering team," or some variation thereof, is ordinarily the most important team in the company. In more traditional firms, internal development departments are all over the place. IT departments are common locations for development teams, but I'm not convinced that software development is part of IT. The skill sets are very different. You don't stick your accountants in the IT department because they use computers, do you? Software engineering may be closer to IT than accounting, but IT focuses on infrastructure (networks, cabling, security system, desktop support, etc.). It would be like lumping your car designers in with civil engineers because they build roads. You don't need to worry about departments and bureaucracy if you're a small company or a startup. Just remember that IT people can't manage software engineers.

Once you work out where you'll put your eager team of code jockeys, you need to hire them. You can do this in all the usual ways, from putting an ad in the paper to posting on job listing sites or just asking around. I've found that developers don't respond extremely well to LinkedIn or Monster job postings. They respond much better to listings on tech-related sites. I've had the most luck with our jobs page on StackOverflow. StackOverflow (SO) is one of the most visited sites for programmers. They get around 12 million unique visits per month from the United States alone. That's a pretty good sample of programmers in one spot. My guess is that SO works well because it signals to potential employees that your company already has an investment in software development, so it might prove to be a good place to work. This is total speculation and anecdotal musing, on top of which there might be some confirmation: I'm a developer who hires other developers, so it might be the kind of place where I would want to get hired! Either way, StackOverflow has worked out well for me in the past.

Recruiters are another potential resource for recruitment. Tech recruiting is big business. Statistics are few and far between on this, but some educated guesses point to between $4 and $8 billion spent on technology-related jobs each year (depending on what you consider technology). If programmers account for just 30 percent of that, it's still big business. This makes the recruiting agencies extremely aggressive, more so than any industry in which I've worked. I don't remember the last time I went to an industry event, tech conference, or even a local programmer meet-up and didn't see a recruiter. These guys show up and bring sandwiches to everything. It's not that I mind them, but they make companies very nervous. The current market is extremely competitive. I recently had a seventeen-year-old intern get a job offer in San Francisco for $130,000 a year. He was still in high school! And 25 percent of $130,000 for a recruitment fee is pretty good for a day's work.

There are a few problems with this. Recruiting companies can start to weave mistrust in tightly knit groups. I have developers who work for me who have fairly large online profiles—they contribute to code projects in their spare time, they write blogs about certain technologies, and help out at programming events. These developers are in the spotlight, so they're highly visible to recruiters. As a result, our office's main line gets between one and five calls from recruiters asking to speak to our employees every single day. The way I deal with this is to pay $50 to anyone who puts a recruiter on speaker phone in the middle of the office. If they decide to go for an interview with a competitor and report back about what they're working on, they get an additional $100. It builds trust among our developers and turns the hunting game into a comedy game. It's also a really cheap way to get intel on what our competitors might be doing! Bringing in employees from a recruiter does a small amount of damage to this type of culture. I'm not saying we never do it, but hiring from within our own resources makes a world of difference.

What to look for

If you want to hire a full-time programmer and you don't have an existing department to manage their recruitment, how can you tell if a coder is any good? This is a tough question, even for a development company like ours. I have a set of rules you can follow that might help spot the right kind of person.

These are the must-have skills for the kind of work that I do. If you find this book useful in any way, then the types of projects you'll want to work on will most likely require the same sets of skills laid out below.

Interpersonal skills

Wow, interpersonal skills at the top of a list of skills for developers? What am I doing? This is nuts! No, not really. The average developer may not be a ball of pure charm, but showing the ability to communicate, be warm, and at least appreciate humor is vital. Engineers are notoriously cranky and introverted. Think of "Nick Burns, Your Company's Computer Guy," the recurring Saturday Night Live character always telling people to "MOVE" out of their chair when he comes to fix a problem. Software teams work on projects bigger than one developer, so being able to respect another person's space, understand what they want, and keep a friendly team atmosphere is vital.

Source Control

Source control is a way of tracking the changes to your code in a single repository (often referred to as a "repo"). This is crucial to a modern developer. It works in a similar way to how "Track Changes" in Microsoft Word works, where one developer can come in and see any changes to the code another developer has made. This lets the programmer see the thinking behind certain problems and solutions. Reading somebody else's code without the context provided by a versioning or source control system is extremely difficult. It's also close to impossible to work in teams without it.

Programmers write code on their laptops or workstations, and then when they've reached a good stopping point, they make "commitments" of their code to the repo. These can be of any size but will have a little comment associated with them describing the purpose of the code. This is a little like saving your Word document to a server. Other developers in your organization (or future developers) can see that "commit" and add to it, comment on it (say whether they think it's good or bad), or make it live on the actual production system.

The two most popular source control technologies are Git and SVN. I personally prefer Git for a number of technical reasons, but if your coder knows how to use either one, they will be able to adapt quickly to the other. There are a number of services that provide GIT and SVN servers, but the biggest are Github and Bitbucket.

A word of caution, here: If a developer tells you they don't use source control because they "don't really need it," do not hire them. I would not hire an agency, freelancer, or a PhD in computer science from MIT if they said that—source control is vital. Don't let anyone tell you that is debatable.

JavaScript

I don't care if they are a C++ superstar or if they contributed to the Ruby core code, if they don't know JavaScript, it's a sign that they are not moving with the times. If they have high and mighty attitudes about "real languages," let them fall on deaf ears. I can't believe JavaScript became what it is now either, but programmers need to get over that. JavaScript is here to stay.

Frameworks

Any modern developer should know at least one or two programming frameworks well. Ideally, they know at least one frontend and one backend framework. If they are exclusively a frontend developer, I would look for Angular or React, which are currently hot. If they are not familiar with a framework, they may make a decent technical project manager, but they are not cut out for a day-to-day development job, especially working in teams.

Backends

If they are a backend or full-stack developer, I like to see a good understanding of databases. Specifically, some SQL experience (see The Anatomy of an Application: Databases for a refresher). NoSQL is cool, but a more senior role would have to get my attention with talk of optimizing databases or something equally nerdy.

A college degree

About 90 percent of the people who work for our company have a college degree. College grads tend to structure their planning and communications slightly better, and those with computer science backgrounds tend to think through code and data architecture design slightly better than those who don't have degrees. These are fairly small differences, and none are deal-breakers. Self-taught programmers are just as good if not better than their college-educated peers, but there may be gaps in their knowledge that they are unaware of due to lack of exposure to broad technological overviews.

Beyond the Basics

These are the little things that might take a potential hire to a slam dunk. They are much more subjective, and harder to quantify, but I've tried to list a few.

Cultural fit

I don't work in the granola, crunchy, bean-bag-chair-filled Silicon Valley, but I do think culture is extremely important. I wouldn't hire anyone I wouldn't want to go out and have a drink with. If you have to work with someone for eight hours a day, five days a week, you had better be able to get along with them. Our company's tent is very big, but we look for smart, nerdy people who like to laugh. The smart and nerdy part is not very hard to find amongst coders, but a sense of humor is a must have to work with me.

Blog

This is just a "nice to have," but it makes me think that a programmer is the kind of person who is trying to be more than just the average coder. If they have a giant portfolio, or a code repository with tons of community project commitments or games they have made or something similar, it shows some dedication to the art. But a blog makes a nice addition. It's something I can read that shows communication skills, dedication, and thoughtfulness. Not having a blog is not a deal breaker, but I like to hire the kind of people who would code for fun, even if that was their day job.

Contributions

Contributions to community projects in a portfolio or a code repository are an added bonus. It doesn't have to be anything huge, but the significance behind it says a lot. It shows that the developer looks at projects beyond their normal scope (that they go above and beyond). It also shows an interest in learning and some level of teamwork, even though they probably don't know the team personally.

Positivity

Very few things will make me fire a person. Still, no one wants to work with an asshole or someone who makes them feel crappy. Programmers are often smart, confident people, which can make them prone to a little arrogance. They carry it around like a chip on their shoulder, and some of them justify being cruel in the name of being "honest." This isn't acceptable. Positivity breeds positivity, and a happy workplace is a productive workplace. Even if you're good at your job, you can't be a grouchy ass.

Hiring Freelancers

But do not ask me where I am going,
As I travel in this limitless world,
Where every step I take is my home. —Dōgen

I recently performed a code audit for a client who hired a developer they met online. During a code audit, one set of developers looks through all the code of another developer and tries to find where there might be errors, security holes, or bad practices. The client, concerned about the slow pace of the project, wanted to find out if another developer could come onboard to help out.

The company had spent around $12,000 a month for the past fourteen months on the project, and it didn't seem to be going anywhere. When we opened it up, we found that the developer had purchased the majority of the code from Code Canyon (https://codecanyon.net/), a site where programmers can purchase code libraries large and small. The developer then spent the better part of a year trying to modify someone else's code to fit the needs of the company's project—and he did an extremely poor job, at that. What could our client do? Try to sue a single person to get back wages they paid him? In the end, they scrapped the project and learned an expensive lesson.

This is obviously a sad, terrible example (and a fairly rare one). I've worked as a freelance programmer, and I know how difficult it is to engage in large projects with little to no backup or support systems. But the example illustrates that there is also a degree of risk in managing freelancers without any technical management on hand.

Freelancers are everywhere in the tech world. This is partly due to the new economy, where everyone is their own boss, and partly due to the nature of software development. The "Lone Wolf" archetype is particularly strong with software developers. I suspect this is due to the fact that for 90 percent of developers, programming is an intensely personal and lonely job. It requires hours of sitting behind a computer screen, thinking intently on a single problem that might take days or weeks to solve. Even getting to that point, learning how to program, can discourage the more socially inclined. This results in a lot of nerds out there who are very good at programming but not exactly brilliant at the soft skills of business. The good news for you is that these people are often for hire, and they are much cheaper than agencies. The only problem is that it means the management of the project, and the programmer, falls on you.

It is hard to find a good lone wolf coder, but you can spot a good one from a high rating and a lot of projects under their belt on a reputable freelancer site. I've been there, both as a freelance coder looking for work, and as a client trying to find good freelancers whom I could hopefully coax into a full-time gig if they worked out. It worked in the past, and I've even hired folks I met as freelancers. I've also had some pretty bad experiences. How can you mitigate your risk?

The best technique I've come up with to make sure I'm hiring a good freelancer is to hire more than one freelancer for the same thing. That may sound a little crazy, or expensive, or both, but it works. I'll take the smallest piece of a project that I think might allow insight into the skills of the programmer, and then I bid that out. I'll then take the three best-looking offers and award the project to all three. When the first portion of the work comes in, I'll be able to tell what kind of person I'm working with. Did they go the extra mile to make something beautiful? Or did they do the bare minimum work to accomplish the task? Did they understand what I was getting at and build something that achieves that goal with elegance? Or are the processes returned convoluted or confusing? How quickly was the work completed compared to the other developers? Finally, how did the quality of work, communication, and cost stack up compared to the other developers?

Sure, this costs me more money in the beginning, but it saves so much money down the line that it's hard to overstate.

Beyond this technique, there are some basics to follow:

- Make sure they have relevant experience in the field and that they can show you some work they've done that is similar to your needs.

- Get a few references. These could be customer ratings on a respected freelancer site like UpWork (http://www.upwork.com) or you could ask for the email addresses of three previous clients.

- Make sure to ask them how they go about development. If they don't mention source control (Git repositories like Github, Bitbucket, or the older SVN), start worrying. See the section on source control in the previous chapter.

Last, when working with someone you don't know very well yet, my advice is to start small. Break the project up into the smallest manageable pieces and assign a new developer the first small milestone. It's much better to lose $1,000 on a small project than to lose $100,000 because you've bought a pile of junk.

Hiring a Digital Agency

"I have a well-deserved reputation for being something of a gadget freak, and am rarely happier than when spending an entire day programming my computer to perform automatically a task that would otherwise take me a good ten seconds to do by hand."—Douglas Adams

Whether you want to augment an in-house team or bring an entirely new product to life, odds are, you're going to entertain the idea of hiring an agency. By an "agency," I mean everything from advertising companies offering digital design services to pure propeller-head software engineering firms. As long as they do "work for hire" jobs, where the client ends up owning the intellectual property and there is more than one person working there, we'll call them an agency.

Agencies come in all shapes and sizes. It's not unusual for an agency to have most of the staff and expertise you'll need to take your application from concept to launch and beyond. They will have programmers, graphic designers, project managers, and even someone to bring you tea while you wait in their lobby. This is the type of organization I run day to day, minus the person to bring you tea. Sorry.

What to look for in a digital agency

Other than the stability a larger organization brings you, there are three main factors to look for in an agency: location, specialization, and reputation.

Location

Having your developers close enough where you can be in their offices within half a day might sound like an old-fashioned idea. In this age of telecommuting and virtual offices, why should you care about where someone works.

Sharing a common geography also means you share a common culture, language, and frame of reference. Odds are, your end users are also in either the same or a similar locale. If a developer were building new ATM software for a bank, it would help tremendously if the design team could spend a few days watching existing customers use the old systems. Nothing drives home a point about user experience more than watching a pensioner struggle with a badly designed interface just so they can withdraw some cash to go shopping.

Empathy breeds good design. It humanizes the code an engineer writes, because now it becomes personal. Teams can instantly see how their creations touch the lives of the very real people who use them. This is especially important in engineering, where most problems are so abstract that concerns become almost entirely practical and programmers forget that actual human beings will interact with their systems. That's hard to do when your development agency is in Canada and your bank is in Mexico. It also helps if you can drive to your developer's offices and yell at them when things don't go right. Okay, okay, it might not help, but it offers some catharsis.

Specialization

Specialization requirements are often the driving factor in a software engineering hiring decision. If you have an entire server environment running Windows and an IT staff trained to support Microsoft technologies, you're unlikely to hire a Linux programmer. Specialties range wildly in the industry. My team focuses on hybridized/cross-platform application development. I know a number of excellent companies that are technology specialists, providing Ruby on Rails experts, Microsoft .NET programmers, or native Apple iOS development only. Some firms focus only on mobile or only on web design. Others will specialize in a certain industry, like healthcare or financial services. The programming world is a big and broad one. If you're a startup in the food service industry looking to build your stack in PHP, there is probably a company out there for you. They might not be close by, but their expertise would outweigh location considerations. Keep in mind that specialists always cost more. If you get to pick the type of work you do and call yourself an expert in your field, that means you can—and certainly will—charge a premium.

Reputation

Portfolios or client testimonials are important regardless of whether you hire a two-woman freelance shop or a 500-person strong New York design firm. One of the problems you might run into with an agency is that the majority of their work might be under NDAs or white-labeled as another firm's property.

Sadly, some of the coolest things my team has ever built will remain a secret forever. If you hire a software firm to build your medical health records system, you don't want to have them advertise it on their website! Even worse, we've built award-winning systems only to watch our clients go and claim their Webby Awards. *Sigh.* We did get a thank-you email, though.

But I digress. Suffice it to say that, due to secrecy, many agencies will show you a list of their previous clients and then give you access to a few references you can call to check up on them.

A word of warning. This is going to sound a little weird, but be careful of the wording you see on agency websites or in their portfolios. Most agencies are going to sell you on their previous work history to establish their expertise, perhaps with a big wall of logos you'll recognize on their website or portfolio page. There is nothing wrong with this . . . at least, I hope not. We have a big wall of logos just like the one I'm describing! However, you need to read between the lines.

If the site says, "We've done work for the following companies," and they show a bunch of logos you recognize, great! Ask them about some of the projects anyway, but it is usually a good sign. On the other hand, you'll see variations of that sentiment that are slightly more weasel-ish. "Our developers have worked with these major brands" is a favorite. Basically, this is the weasel's way of saying, "Well, our company hasn't worked with any of these companies, but some of the people who work here used to work for companies that worked for those companies, or they worked for those companies themselves at some point." Their sentiment is technically accurate misleading.

If I had a job in college making coffee at Starbucks, I shouldn't be able to put their logo up on my website for financial services consulting because I "have worked with this major brand," yet this is unbelievably common. It is more common among ad agencies and marketing firms than it is in pure software development shops. The reason may be that ad executives run these agencies. Ad execs see themselves as their product. If they worked at Nike or at a firm that worked for Nike, they feel the experience they gained is valuable and worthy of promotion in their new endeavors. Software development shops are generally run by "reformed" programmers. They might not be able to take the "creative leap" more common on Madison Ave.

Offshore Development

"I've decided to take advantage of outsourcing. My next novel will be written by a couple of guys in Bangalore, India." —Tom Robbins

Ah, my least favorite model. This isn't because of an inherent bias against any particular country or region; it's because I've learned hard lessons time and time again.

Not a week that goes by that an Indian development team doesn't call our offices looking to partner with our company to help "augment our team." It always looks like a great deal. You can hire an experienced fill-in-the-blank technology programmer who will be completely handled by an English-speaking project manager and will cost half of what we pay our U.S. programmers. So, what's the problem?

Cultural Issues

When I first started out as a lone freelancer, years before my first technical hire, I dabbled in offshoring. It was a great way for a young company to be able to tackle projects that I just didn't have the time for or that I needed help with. I tried development shops and freelancers based in India, Pakistan, China, Ukraine, Russia, Egypt, Poland, and even Peru. The results varied but were never great. I experienced better results with a few specialist development partners in the United Kingdom and Australia, and the fact that I achieved better results with people who had more in common with me made me feel somewhere between guilty and angry. I was missing the point.

I spent most of the last twenty years of my professional software development career building applications for American businesses. Americans have a particular set of cultural norms and expectations surrounding business and product development. The reason I found it easier to work with Australians, who have a twelve-hour time difference, than with Ukrainians, who only have a six-hour time difference, is purely a cultural problem. Australia and the United Kingdom have sets of norms, expectations, and communication styles similar to my own—not to mention a common language (if you consider Australian English a common language!).

When you ask a programmer from China to do something as simple as "add a button to the bottom of the box and label it 'OK'," that is what they will do. They will create a button, put the text "OK," on it, and call it a day. But to a developer from the West, the word "button" carries a lot of weight, even subconsciously. It means something that, when a user clicks on it, will act as though it is depressed, then released, changing color and state as it transitions. That's how all buttons work in Australian apps, so why do anything different? When the American project manager complains to the developer from China that the button doesn't act "button-ish" enough, the programmer assumes that Americans must be insane. The entire experience creates tiny hostilities that lead to the developers becoming less and less likely to compromise on anything, and the project managers accusing the programmers of being lazy.

When you don't share a common language, let alone a common design and aesthetic reference, you cannot have a productive relationship. Cultural understanding goes both ways. It's neither the offshore developer's fault nor the local project manager's.

There are ways around this. More experienced offshore developers will give the customer what they expect, rather than what they asked for—but it is difficult to know what you'll get up front. The other way to mitigate these issues is to have incredibly detailed specifications. Nevertheless, building scopes that detail how every button is to react takes so much time that the cost benefits from offshoring no longer make sense.

When you ask a developer to build a particular feature, there is no specification clear enough to outline everything all the way down to the methodology of doing it. If there were, you might as well write the code yourself—not to mention that your micromanagement would drive a programmer out of her mind. Why would anyone bother becoming an expert in developing mobile apps if their opinion and creativity were not required?

When there isn't an exactly defined item in a specification, a coder uses their own judgment and expertise to decide how to best solve the problem. The invisible decisions between every functional outcome are constantly flavored by a programmer's background. My experience is that you will be happiest finding developers with backgrounds similar to yours.

The same is true of your users. If a programmer has never met anyone like your intended user, how are they going to add any value other than being a code monkey? If you design ATM software, and you don't take your development team to the bank using it, you're in for trouble. Seeing the little old lady struggle with the font size or trying to decipher whether or not the transaction is complete makes the goal personal. It gives it more meaning to the team building the software; they can imagine their grandmother using the application they're writing, and they can feel the weight of the importance of what they're working on. As a result, they work harder, speak out more when they see something going wrong, and feel personally invested in the project. It's almost impossible to do that if you live in a country where that bank doesn't do business.

Managing Offshore Teams

One of the biggest challenges when it comes to managing offshore teams is the distance. Finding the time to meet, review progress, and live-test new features becomes a burden to both sides of the team.

There are occasions when this is a benefit, especially if you use offshore teams to augment local developers. In this scenario, your local developers review code created overnight by the offshore team. Your QA staff can arrive in the morning with new features to test and can have feedback for the development team when they walk in the following morning. This doesn't mitigate the frustration caused by not being able to reach anyone on a Friday afternoon when you find a critical bug in your freshly launched system. It also often means meetings late on Sunday evenings.

Offshore Costs

Costs depend on the location of your offshoring. Outsourcing to countries with a much lower cost of living results in large cost savings. But due to all the issues mentioned above, offshoring often ends up costing you more in terms of hiring experts to fix poorly programmed systems or having to completely scrap and rewrite failed projects.

If cost is not an issue, the benefit to offshoring is the massively increased pool of experts you have to choose from. You may not be able to find someone in your area, or even in the entire country, that has the experience with an arcane technology you must integrate with for your project. On the other hand, if you face a large amount of simple yet tedious tasks, then the cost benefits of hiring a foreign development team might free up resources and create some wiggle room in the budget for your local developers. Having a junior, $10/hour Indian developer programming hundreds of simple scripts for scraping data off websites might permit your $150/hour local developer to focus on the parts of your system that demand the most attention.

Typically, offshore teams charge on a per-resource model. You pay for a programmer's full-time participation in a project month to month, even if you don't use all of their time. While still cheaper than hiring a part-time freelancer from your area, if you only need twenty hours a week of work for the next three months, finding an offshore company to provide the cost savings you are looking for may prove difficult.

What to Look For

If you are dead set on hiring an offshore firm, look closely for three key items.

1. Do they have previous experience in projects similar to yours? If possible, get references from the company that you can call for firsthand feedback about how the team helped on a similar system. This can occasionally lead to you calling a competitor. I find that a white lie about who you work for provides honest assessments, as well as the occasional bit of business intel.

2. Make sure you have solid contracts with clear intellectual property ownership laid out, as well as solid non-disclosure agreements.

3. Many large offshore organizations have locally based project managers that act as points of contact. These resources are favorable because they likely speak the same native language as the developer with whom you are working.

CHAPTER 7

WHAT DOES IT COST?

You know what you want to build. You have a few rough ideas jotted down, and you're ready to find a development partner. There is one big question outstanding: What is this all going to cost?

This is one of those "How long is a piece of string?" questions. It's very difficult to say how much an unknown project costs. In the past, our development shop built large web applications in excess of a million dollars and small apps for under $20,000, but there are hundreds of small, mitigating factors.

A good rule of thumb is that application costs are similar to buying a car. You can get something with four tires, a seat, and an engine for a couple thousand bucks, but you're not going to want to drive from the Cape to Cairo in it. By hiring a good development agency, a typical tech startup can get to a "minimally viable product" with about $50k to $100k in development depending on the complexity. A larger scale, enterprise-grade system is typically in the $200k to $500k range and could cost much more. These may seem like high numbers, but if you consider the cost of starting a traditional brick-and-mortar store or hiring an internal team of developers to work on a product for a year, they're significantly cheaper.

These are very general ballpark figures, and the costs vary wildly depending on the type of company you choose. These are also the cost estimates for the initial development. This doesn't take into account marketing, training, hosting, or maintenance.

You Get What You Pay For

For some reason, this is a lesson we all forget when presented with true costs. Everyone wants to get a good deal. Certain professional services are very easy for most business people to estimate. If you need a small business accountant, you know you're looking at $60 to $150 an hour. If you're looking for a big, national firm, you might be looking at closer to $400 on the high end. Lawyers vary from $200 an hour on the lower end to $600 and more at bigger firms. We intuitively understand these costs because we realize lawyers went to school for a long time to get where they are, and we have read enough contracts to know that we sure don't want to write them. We also know that if something goes wrong, we want our lawyer to know what they're doing!

In contrast, the average manager hasn't written a whole lot of code in their lifetime. We're constantly presented with teenagers on the news creating new products that hit it big. On top of that, you're probably inundated with email offers from Indian development firms offering to program for $15 an hour, or you see freelancing sites advertising entire apps built for a couple hundred bucks. On the other hand, you might receive a quote of a million dollars for an app from a large agency in New York or Los Angeles. It's not surprising that people don't know what they should pay a programmer! Never fear, I'll go into more detail about my personal experience with costs in the next section.

The thing to realize here is that you get what you pay for. If a U.S. developer charges $150 per hour, they are probably worth it. There are ways to work out how good a programmer, or an entire development team, is (see Finding Developers), but setting your expectations early is important.

When you hire a developer or team, you're not just building software for today. If you take your business seriously, you will build something that lasts seven to ten years. That might not sound like a long time, but it is an eternity in the tech world. With this in mind, you need a development team you think will stick around to work on your code. You need to make sure that the language they're programming in is also going to be there in case you have to replace your original team. Finding someone who can code in an esoteric, fly-by-night language is going to be very difficult ten years from now.

Code quality is something to keep in mind, too. When you hire a cheap developer, you don't really have any guarantees that they do things "the right way." Good programmers adhere to development standards that make it easier for others to read and edit their code. If you build something the wrong way in the beginning, it might be impossible to replace that programmer down the line. Bad code is also slower, less secure, and harder to manage, meaning it's much more difficult to extend the application's functionality.

o sum up, you probably don't need to spend $300 an hour to find a great programmer or team, but you really don't want to hire anyone who calls themselves an engineer at $20 an hour.

Estimating Development Costs

Fixed Bids Vs. Time and Materials

Most developers prefer to work on a time and materials (T&M) basis. Development estimates are notoriously unreliable, even from the best programmers, so doing T&M protects the programmer from ending up in a situation where they pay to work. Some freelancers and agencies offer fixed bid contracts for smaller phases of work. Be wary of a developer who gives you a quote for a giant project—they've either padded it so much that it's probably unreasonable, or you're going to get sub-par work because they're trying to get the minimum work done in the minimum amount of time.

Our agency has a number of hybrid models. They offer fixed bids for the first phase of a project and then estimates for future phases (getting vaguer as the phases project into the future). This is because 99 percent of the time, a project changes goals and features as it progresses. The other model they use is a retainer-based time and materials (T&M) model. This takes a monthly amount of hours allocated to the client. Costs decrease the more hours a client decides to dedicate monthly to the project. This is often referred to by agencies as the "discount schedule."

Freelance Coders vs. Agencies

If you're looking to build something small, or doing a one-off project, you might be completely happy with a freelancer. You can find thousands of freelancers on sites like Freelancer.com or Elance.com. You'll find developers of skills and costs all over the board, from $10 to $300 an hour. Granted, the high-end costs are mainly for specialized services, exotic languages, or experts in a particular field. Normally, you write your own scope and then publish it for the community to bid on. From there, you can peruse the profiles and the portfolios of your potential freelancers. On a site like Upwork, the average hourly cost for a developer is around $20 an hour. This is mainly due to a huge number of low-cost developers in India, Pakistan, China, and Eastern Europe willing to work for much less than the average U.S.-based programmer (generally $40 to $100 per hour on Upwork).

Dealing with an agency is a little different. An agency has to hire project managers, developers, and designers, as well as pay rent, accountants, lawyers, and the usual slew of business expenses a freelancer working from home doesn't have to worry about. As a result, they're going to be more expensive. You get better code and a much more professional attitude from an agency because they have a reputation to uphold, and they have to make payroll! Typical costs for a specialized development studio are around $100 to $150 an hour. This increases in large metro areas and on the West Coast, where the second tech boom led to massive inflation. San Francisco and Silicon Valley produced a talent void gobbling up programmers from all over the world, driving up the costs of freelancers and agencies alike, especially in those areas. Similar hits took place in New York City, Chicago, Austin, and Boston, though not as hard.

Estimating Total Costs

It's impossible for me to tell you what your application will cost to build without knowing about it in detail. This applies to every developer out there, so be wary of programmers or agencies who don't start the conversation with a discovery or planning phase. I would suggest you plan on spending 10 to 15 percent of your budget on planning, scoping, and wireframing throughout the project (with most of that up front). It's always better if you split up your development into manageable phases, each one with its own planning and prototyping phase done before the coding gets going, but that's not always possible.

Spend another 10 to 20 percent of your budget on QA, testing, and bug fixes. If your application is mission critical in any way, then you might want to increase that percentage.

The remainder of your budget will go toward design and development. The mix of cost here depends on the type of application you're building. A game, for example, might require many more level designers and graphic artists than programmers. A tool for your insurance agents to generate quotes for customers might not need more than one designer for a few days.

Technology Factors

Cost-wise, no technology is equal. Some wide generalizations are about to follow, but they should be good enough to give you an idea.

1. Open-source technologies are cheaper than propriety solutions. This means that, in general, you'll find applications that live on Linux servers cheaper to develop than those that live on Windows servers. Oracle is much more expensive than PostgreSQL. Yes, there will usually be tradeoffs somewhere (an open-source technology may be less professionally supported), but the gap is smaller than you would think.

2. Hybrid or cross-platform technologies run cheaper than native ones—this is because you are using repurposed web technologies. These technologies are more widely accessible to novice developers, so you will find more people who understand the technology.

3. Web is cheaper than mobile (that includes the mobile web).

4. Building the Windows way is typically more expensive. By this, I mean .NET technologies, Windows servers, and so on. Open-source servers and technologies are free, and as a result, easy to find. This leads to more developers in the market, driving down the prices. This may change soon. Microsoft is open-sourcing many of their development technologies. Eventually, they may give it all away in an attempt to woo programmers.

Don't Forget!

As with everything in life, there are often hidden costs. Of course, these shouldn't remain hidden, and your developer or project manager should discuss them with you before beginning.

Licenses

Developers don't build everything from scratch these days. There are great libraries of code out there for making everything from beautiful graphs and charts to starter kits for entire mobile apps. Almost every one of these building blocks that programmers "pull off the shelf" comes with some sort of license.

Commercial licenses come with clear pricing structures. They might bill you based on how many systems you're going to install their software on, how many developers you have, or how many users you intend to support.

Open-source licenses can be a little trickier. It depends on what you're planning on doing with your software, but sometimes they're free and sometimes they can cost extra. If you decided to go with a popular database like MySQL, that's great. The MySQL license costs money under certain specified conditions. These are mainly if you package your application to sell as a whole, like on a disk, or if the entire thing is made available for download. In the case where you use MySQL on your server and other users access your application via the web, you're probably free to use it without cost.

Additional Costs

In estimating any costs, here are a few things you might forget about.

Servers and Hosting

Most software systems these days run on the client-server model. The server side does the heavy lifting, and the user has some sort of intelligent interface to the application (the client), like a web browser or a mobile app. It doesn't cost too much to get a mobile app in one of the app stores (Apple charges between $99 and $299 a year, depending on your license), but it does cost money to have a server up and running somewhere.

Your typical costs for a basic app using a server somewhere look like this:

- $10 a year for a domain name
- $200 a year for an SSL encryption certificate
- $20 a year for a dedicated IP address (needed for the SSL certificate)
- $100 a month for a server (including server software)

That architecture would work for a typical project, utilizing open-source software. If you plan on scaling to much larger audiences, that server hosting cost grows rapidly. It depends drastically on what type of data you store as well. Storing video requires much more server power and disk space than an app that just stores text records. Also keep in mind any regulatory requirements you face. If your app needs to be HIPAA compliant, expect your costs to double or triple.

All in all, your server hosting costs are a factor of how much data your users need and how much storage they require. A good rule of thumb is your per-user hosting cost should be below $0.25 per month per user for a simple text-based system and under $2 for a video-based system.

Maintenance and Support

One of the largest costs encountered in development is continuing support and maintenance. If you hire a development firm, you will want to keep them around to help you make changes, fix bugs, and keep your system healthy. You don't want to get into a situation where you call your developer and ask them for something, but they're too busy working on their next project to help out. This is where retainers and maintenance contracts come in. You will want to budget around 15 percent of the total build-cost annually for support. That does not include new features, but it allows you to support your system past the launch date. Be extremely wary of programmers who don't provide you with an ongoing maintenance plan—it's a good sign that they're a fly-by-night firm, or they're too desperate moving on to the next project to support your platform into the future.

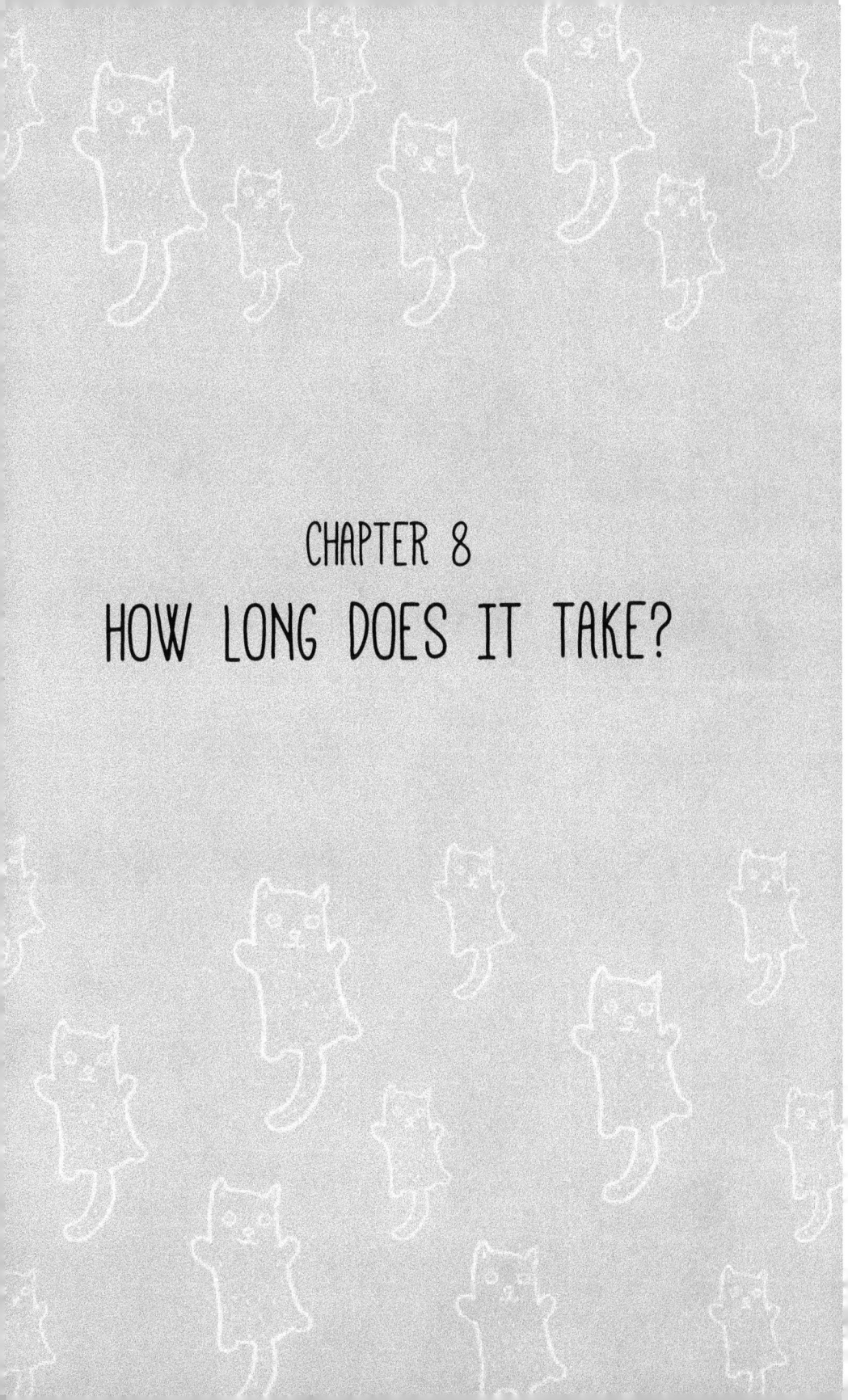

CHAPTER 8
HOW LONG DOES IT TAKE?

Another "piece of string" question is, "How long on average will an app take to build?" While nearly impossible to answer without knowing the specifics of a particular project, here are some general guidelines I can offer.

- The most basic of applications (a simple one-page website or an extremely simple mobile app) takes less than four weeks to build.

- Moderately complex applications (a standard e-commerce system, a client portal with users and logins, an interactive app with animations and in-app purchases) take between three and six months.

- A complex application (one that interacts with many third-party systems, uses any type of machine learning or intelligence, includes payment systems, contains a lot of content, and is mission critical) can take over a year to develop.

If you absolutely need an answer, a recent poll of software development companies reported the average time to build version 1.0 of an app was eighteen weeks. That does not include the planning and requirements gathering phases, which add an additional two to four weeks depending on the size of the project.

How Long Does It Take to Build an Application

Software as a Process, Not a Product

Software does not live in a vacuum. It lives alongside other software and hardware systems, and it lives in the user's environment. As a result, software needs continuous maintenance, updates, and care to stay secure, competitive, and relevant. Unlike most products taken to market, software requires significantly more frequent updates. A hairbrush design might not look "old-fashioned" for twenty years. A software interface older than three years might appear dated. This means the question of "How long does it take to build X?" is even trickier. A better way of asking would involve three separate questions:

1. How long would it take to get to version one of my product?

2. What are my ongoing maintenance, hosting, and support plans for the product once it's launched?

3. How many features will I add down the line?

These three questions can help shape your planning and budgeting, not just for the first version of your application but for the future of the product as a whole. In the following pages, we'll look briefly at some of the factors that influence the future of a piece of software. Try to keep in mind that you're building a process, not a product. One that will need love, care, and technical maintenance for the life of the product.

Hardware and Software

All software lives on some sort of hardware system. If you're building a typical mobile app, you need to consider three distinct factors that change over time:

1. **The mobile device.** Smartphone companies release new devices at least once or twice a year to stay competitive. This means new screen sizes and hardware features. We design most apps specifically for certain screen sizes and resolutions. When a new phone comes out with a bigger screen or higher resolution, your app may look odd or blurry on the newer systems. When fingerprint ID hardware appeared on phones in 2014, users loved not having to remember all their passwords. However, any app wanting to use this new feature needed an update.

2. **The operating system.** Not only do the hardware devices update, but the underlying software of their operating systems do as well. These bring new tools and features to the phones. Some of these updates eventually break existing features in your app. It happens all the time. Your developers should receive advanced copies of the new operating systems to test your existing software. You must make sure they do this, and you need to pay them to do the work.

3. **The server.** Almost every app out there relies to some extent on a server connected to the Internet. Whether it keeps track of scores for your game or saves your user's login credentials, apps rely on servers. The problem is that servers have software and operating systems, too. They require security updates, virus scanning, and general maintenance just like any computer. Someone needs to do this work.

Other factors to keep in mind are things like web browsers. Chrome, Firefox, and especially Microsoft's Explorer release updates almost monthly.

Third-Party Systems

Your software will almost certainly not be completely self-contained. Developers use code libraries to make their lives easier. Code libraries get updates, overhauls, and even discontinued.

APIs are another consideration. When Google discontinued their Maps Engine (a tool for collaboratively working on maps) in January of 2016, it caused tens of thousands of web apps to break. This happens frequently, so someone has to monitor the status of the APIs you're going to use.

Culture and Fashion

Finally, we have to consider the user's expectations of your software. Remember a few years ago when every website or app had buttons and boxes with nice rounded corners? You don't really see those anymore, do you? This was great news for designers and developers, because they were a pain to work with, but that's not why they disappeared. They simply fell out of fashion.

Fashion exists in design the same way it exists everywhere else. We live in a culture, and culture is invisible unless you're looking hard for it. It's as strange to most people to think about their own culture as it would be to ask a fish how the water feels. What water? Icons, symbolism, color choices, and thousands of other invisible cultural and fashion forces change around us all the time. You probably won't know what changed, but one day, you'll look at your software and think, "Wow, this looks dated." Keeping your look and feel up to date is a constant battle but one that is worth it in a competitive landscape.

CHAPTER 9
THE BUILD PROCESS

"Programming today is a race between software engineers striving to build bigger and better idiot-proof programs, and the Universe trying to produce bigger and better idiots. So far, the Universe is winning." —Rich Cook

You've done most of the thinking for your application. You've gone out and hired a great team of programmers to get it built for you. You're an expert in what you can and cannot accomplish for what price. Now comes the tricky bit: actually building the thing!

In this section, I'll discuss how to take your ideas and get them down on paper. This way, you will have something to hand over to your programmers so they will know what to build. We'll also discuss the architectural side of the planning process, how the system will flow, and what screens will look like to your users.

Next, we'll discuss some of the processes encountered in the build process. I won't get into every project management and programming theory out there, but we'll survey the landscape.

Finally, we'll discuss testing your application. This is the crucial step before launch where you'll find bugs and come up with features for your next version.

The Development Process

"If debugging is the process of removing software bugs, then programming must be the process of putting them in." —Edsger Dijkstra (Winner of the 1972 Turing Award)

The "software development cycle" is a common explanation of the sausage making that is software development. It is a fairly opaque cycle, consisting of the following phases.

Planning

Planning includes requirements gathering, needs analysis, scoping, wireframing, and everything in between.

Design

This phase is a catch-all for everything from database design to the user experience (UX). It includes easily understandable user interface design (UI), graphic design, color choices, and more esoteric "design" choices such as code structure and API design.

Build, Development, or Implementation

My favorite of the partially-obvious, partially-meaningless "cycle"—the coding part. This is where developers sit in front of big monitors and do stuff with text that makes other stuff happen.

QA, Testing, or Implementation

Once we code something, we test it. When we find bugs, we fix them. Once again, this is straightforward. No need to bore you with how it all works.

Maintenance

The most opaque of all the steps! Maintenance can mean a lot of different things to different teams, but it relays intention of doing updates to the system once live. Developers assume things will never get to this point. Software is continuously evolving, systems are continuously improved, and so maintenance is rarely ever planned.

Once you complete a full cycle, you have some sort of first-phase product ready to go back to a planning stage for the next phase to improve the product. The problem is, I have never seen a software project work this way, ever.

The very second that the results of any project are visible, the specifications start to change. There is no amount of planning on the planet to account for this. Imagine trying to develop a simple trivia app for the iPhone. You might start with great wireframes, move on to a great-looking design, and then start coding. The moment you play the first version of the game, you notice certain things that need improvement. Perhaps there is an extra step between rounds where the user taps a button saying they're ready. After playing the game a few times, you realize how annoying that button is and want to remove it. That creates the problem of how to indicate the start of a new round to the user but is easily solved by a quick "Next Round!" popup at the top of the screen.

So what do you do? Do you go back to the designer and ask them to mock this up? Do you go directly to the programmer or project manager and ask them to make this change? What is the amount of required planning? Does the change introduce a new flow in the user experience in a way that it seems out of place? And does that mean you should go back and reassess all the buttons that do similar tasks and bring them in line with the new, better process? What happens to the schedule of the project if this happens over and over again?

This is an extremely basic example, but it does illustrate the rabbit hole you can go down when it comes to software projects. Don't panic. There is a solution. It is both a management and a psychological solution. This is the acceptance of software being a process, rather than a product. It is the assumption and acceptance of change, rather than the mindset that change equals chaos. In the following sections, we outline some of the various project management methodologies best suited to building software products. We'll go through some of those steps in the cycle to arm you with the knowledge you need to deal with change and jargon-spewing project managers!

Project Management

"Of all the things I've done, the most vital is coordinating the talents of those who work for us and pointing them towards a certain goal."—Walt Disney

Every professional developer or development team has a strategy for managing their projects—whether they know it or not! Your team should clearly articulate this methodology to you. If not, warning bells should go off. I'm not fanatical about which method is correct. There are people out there who are borderline religious about their methodologies, but like anything, there is never one size that fits all. The only truly correct approach is to have the ability to take the bits that work for you, or for a particular project, and blend them to suit your unique needs in the moment.

Cowboy Coding

Before we get to the two most common project management methodologies, let's examine what it looks like to not have one. "Cowboy coding" isn't as cool as it sounds. This is a derogatory term for a developer who, when given a project to work on, starts coding from day one. This is extremely common with many freelance programmers, usually those working alone.

On the wide open plains of the Internet, a cowboy coder roams free, programming in whatever style they prefer and not worrying about the future. This doesn't mean he or she is a bad programmer. Lone-gun developers can often be quite talented coders. The problem is that they don't have a clear system for communication. Communication with who? Well, everyone.

Communication is a catch-all term these days. "Effective communication is paramount to our success as a team" might be the type of silly thing you find in a Kool-Aid-drinking middle manager's email footer. That doesn't mean it's wrong. Not having a communication plan causes real-life problems.

1. Code is rarely developed in isolation. A project of any significant size requires multiple people, whether at the beginning or during the project's life cycle. If good communication and architectural habits haven't formed early on, it's a sure thing that the system will need a rewrite in the future.

2. Programmers are rarely the end users of the software they write. This means you need to discuss and plan multiple decisions so that your programmer can implement them in the correct way. Having a constant and open dialogue with the stakeholders of a system is extremely important as a result. If your developer is making these decisions on their own, there is no telling what will need to be redone in later iterations of the software.

3. Because there isn't a larger, overarching plan for the product development, you are often left with "spaghetti code." This is a term to describe code that is all over the place, doing only one specific task. Good coders, following strong architectural practices, write modules of code reusable throughout the system. This makes it easier for a new programmer to jump in to help, or to maintain down the line.

One caveat here is that when you meet a really good programmer or team of programmers, they can sometimes appear not to have a plan. This is because after taking many projects through to success, the practices, architecture, and planning are so internalized that they just "do it."

Waterfall

The waterfall methodology was first practiced with intent in the 1970s and is still going strong. It is the project management system most familiar to anyone in business, construction, and government. Waterfall is the standard "sequential" process, meaning there are a number of steps planned out in advance, and someone executes them. This looks something like:

1. Gather the requirements

2. Design the way the system will work

3. Implement the software (by writing the code)

4. Verify that it all works with testing

5. Maintain the system and fix things that break

From the software development side, the planning of what to build and when to build it is frequently plotted out on a Gantt chart. These are those scary looking sheets of lines and arrows showing:

- The duration of each step
- Resources (people) required per step
- Which steps are dependent on which previous steps

Waterfall was the dominating methodology of developing software for close to forty years. Its focus on time frames, budgets, and resources attracted large companies who could afford to develop their own software. Waterfall is still the methodology of choice in most industries to this day. Everyone from construction firms to movie producers use detailed planning, Gantt charts, and resource planning to try and keep their projects on budget and on time.

When Waterfall is Useful

Waterfall is effective if you have an extremely time-restricted development cycle but your requirements are very well defined. When the project plan is at least somewhat close to accurate, waterfall is very good at predicting budgets.

Where Waterfall Falls Short

Because waterfall emphasizes one step after the next, going back to a step to fix or change something is often impossible. It also keeps the programmer's mind fixed on the schedule rather than the quality of what they're working on.

Agile Development

In the early 2000s, the state of software development changed rapidly. Software engineering was no longer an esoteric profession but the profession that built the modern era. New programming languages, education, tools, and cultural importance drove down the costs of developing software. At the same time, software was the new currency in business, so everyone was trying to build the next Microsoft. The paradigm shifted from "building what you need" to "build something we can get to market as quickly as possible." Startup culture took on a near fanatical mindset of giving the customer what they want. This meant building software in small chunks that you could then show to your customer and ask for their feedback. Once you got the feedback, you went back and iterated on the product. Lather, rinse, repeat until you have a perfectly happy customer.

Dovetailing with the customer-centric model was the simple fact that software projects grew larger and more complicated. Outcomes were more and more unpredictable. In order to deal with this unpredictability, agile development, or agile project management, became the new norm.

Agile refers to a number of practices:

1. **Iterative design and development.** Users complete work in small batches. Planning, then execution, testing, and feedback are all completed in short cycles, often called "sprints." These can be as short as a week.

2. **Deliver products as early as possible and continuously improve on them.** The idea is to accommodate changes to the project, even in late stages of development. This continuous improvement and frequent update philosophy extends to the maintenance of the product as well.

3. **Constant communication.** Agile emphasizes frequent communication between the project managers, the people doing the work, and the stakeholders.

There are various Agile frameworks, or sects, which emphasize different approaches to achieving these goals. Scrum might be the most famous of these frameworks, but there are many more.

When Agile is Useful

Project owners, clients, and stakeholders often don't know exactly what they want. Agile lets them see small increments of work, which helps direct the next phase of development. This is invaluable when the scope of work is not terribly well-defined. It also permits good programmers to be more creative and to solve problems on their own.

Where Agile Falls Short

If there isn't a solid plan, agile is a sure-fire way to go completely off the rails. This can be due to a weak project manager not keeping the focus of their clients and programmers. Over-focus on iterative design leads to long project overruns in time and cost alike.

The Blended Approach

Some organizations treat their project management and development methodologies as religion. If the training manual outlines a particular step, that is law. The truth is that there isn't a perfect path for any team or any project out there. Understanding these methodologies is beneficial, but being able to pick what works best for you is what counts.

In my experience, the best approach is a blend of traditional methodologies and agile development. I like to start off a project with a lot of planning, timeline estimates, resource scheduling diagrams, and as much documentation as I can get. While doing this, I try to remember that we'll throw away almost all of my plans at some point. Do as much of this work at the beginning as possible. It allows goals to be set, the creation of a vision, and a clear direction for programmers who have to start somewhere. My line in the sand regarding where we switch from waterfall to agile is after I am able to show a client an early prototype. This allots for a head start on something tangible and then provides the flexibility to adapt to changing requirements.

Planning

"It takes as much energy to wish as it does to plan." —Eleanor Roosevelt

You have your idea. You've worked out how your application will change the world or help your company or make a bazillion dollars. Now the real work begins. You're ready to start coding. But wait! There is more fun ahead. First, you need to plan the whole thing out.

Nobody likes planning. Not detailed planning, anyway. It's time-consuming, requires real rigor, and doesn't produce anything fun to play with. However, it is the most important phase in the development process by far, especially if you're dealing with contract engineers.

What we cover in this chapter is a general survey. Read everything you can about software planning before engaging with a developer. **This is the first and last part of the development process that you have ultimate control over.** On the bright side, you can do this part yourself if you are so inclined. All the tools required to get it done are mostly nontechnical and intuitive. Still, I recommend you work collaboratively with your developers to get it done—but a head start on the work will help you think through the tricky parts of the application and most likely save you money when it comes time to engage with people who charge for their services.

I cannot stress how difficult it is to get the ideas and vision out of your brain and into the brain of a programmer. It's not because developers are difficult people (although they often are), or that product managers are bad communicators (most of the time they're really not). But there is an invisible cultural divide between the two. Programmers work in a world of abstractions and concretes. What you need to get across to them are the concrete essentials and let them handle the abstracts. Telling an engineer, "I want you to build me a chair," probably makes perfect sense to you, but when the engineer brings you a three-legged stool with one leg shorter than the other because they thought it would be better for tilting you toward your desk, you're probably going to want your money back.

On the flip side, you want to be flexible to a certain extent. Cost considerations are one issue. I'm reminded of the old adage, "I can do it quick. I can do it well.

I can do it cheap. Pick any two." That applies here. In terms of planning, your feature decisions directly affect the time, quality, and cost of the entire project. Since, as we have discussed, software is a process and not a product, you can constantly shape and adjust it for the life of the end product.

The first decision you need to make when approaching your initial plan is, "What do I want to get out of this first version?" Do not take this step lightly. Most purchasing decisions you've ever made when having something custom-built don't work this way.

Remember our house-building example from the introduction? Well, no one would hire a construction firm to build a house with a kitchen and a bedroom—with the plan of adding a bathroom later. You'd be setting yourself up for a nonfunctional (not to mention uncomfortable) first few months of home ownership. Most software engineers don't think of software development with the same logic. Probably because they're not often brought in during the conceptual phase of a product. Making your development team aware of the final project goals will help them make better decisions about how the software is architected. This will save you cost overruns and code rewrites down the line.

Unfortunately, most contract development companies won't give you direct access to their developers. They don't do this for a number of reasons, including (1) companies don't trust their developers to have the people skills to interact with paying customers, (2) companies don't trust their clients not to poach their valuable programmers by hiring them, and (3) a lot of developers don't want to work with clients—they enjoy their protected environment. If you're not working directly with a developer, with any luck you can work closely with a technical project manager who can navigate with one foot in each world.

Even if you do find a good company or single developer to work with, that does not let you off the hook of the planning process. Most development companies these days start a project with a detailed needs analysis and scoping process. If they don't, that's a red flag. As little as anyone likes spending money, you want to find a company that charges you for this. I know that sounds crazy, but you wouldn't want to pay a construction company to build an office building for you without first hiring an architect. In a similar fashion, you want to find a development firm that takes planning as seriously.

A good development firm (or single developer, for that matter) guides you through at least two rough planning phases. These are very general areas, and the bigger the company, the more formalized these will be, but you should at least expect a scoping phase and a wireframing phase (some developers roll these together).

Scoping and Requirements Gathering

Scoping refers to the process of working out the what and why of a product in writing. For example, a very small scope might look something like this.

Right now, we have 200 salespeople out in the field. It is very difficult to tell who is doing what at any given time. We don't know if our high performers are in areas with greater needs or if our lower performers just slack off.

What we would like is an Android phone app, running on Android 5.0 and higher, that sales associates log into to see their appointments (pulled from our SalesForce CRM) and log notes about their appointments.

We would like these notes stored in SalesForce as well as a management tool (probably a web application) that records our salespeople's GPS locations and the time so we can monitor their routes.

We would also like this app to generate reports showing the time spent in commute, the sales that actually came from the visits, and the average visits per day ranked against the distance traveled.

Don't try to write a technical document. Even if you're an engineer or have some programming experience, technical-jargon-laced verbiage is the quickest way to confuse a programmer. I know that sounds counterintuitive, but the more descriptive, the better. Keep in mind that phrases like "end-to-end" or "user-centric" can mean different things to different people (at best), or nothing (at worst, and probably more likely). If you need to write something highly technical, like API documentation or network infrastructure, put it in an appendix.

While this is a very basic scope, it's not terrible, considering I made it up on the spot. It details the important information:

1. What we want to build
2. Why we want to build it
3. What it will do
4. Who will use it
5. What platforms it will run on

The "why" question is more important than you might think. While it might be obvious to someone building a house that four real people are going to sleep, shower, cook, and entertain in this house, the basics of your project might not be as obvious to someone who doesn't know how your specific business functions to work out who will use this software and why.

A developer and project manager can almost always make better decisions if they have context. Your programmers will not inherently understand your business, processes, or customers. Technology is a way of enabling business and rarely is the business itself. That means it's highly unlikely that the developer you hire for your custom animal slippers business knows anything about fluffy rabbit shoes—or has ever bought slippers at all. In fact, one of my developers wore the same pair of shoes for over ten years . . . only one pair! But I really digress.

The takeaway is that painting a clear idea of what you want to do and planting it in your developer's mind is almost as important as the next part of scoping: requirements gathering.

Requirements gathering is a term for "writing down all the stuff the application has to do." This means a feature list. In general, you start with a long laundry list of features, both obvious and non-obvious. Don't leave it up to chance. Just because you think it might be clear that a user has to create an account on your system to interact with it, doesn't mean a developer does. For our sample application above, a small section of our scope might look something like this.

Management Portal

The super admin can create managers and assign them a group of sales staff to manage. These managers can only see the reports of their sales staff.

Managers can log in with a username and password. From there, they can see a menu with the following choices:

Reports - Where they can choose reports to download or view

Staff - A list of all their sales staff (they can click on each staff member to drill down into their sales stats)

Contact Clients - A section where managers can send one of three preprogrammed polls to all of their sales staff's leads to request a quick performance evaluation

There are other parts to this phase you might see as well. "User stories" are a big one that dev firms like. These are descriptions of what a user might see or do under certain circumstances. "Permissions documents" is another common scope document, which lays out what different types of users can do (i.e., admin users can do everything; editors can edit, create, and publish content; authors can create content but not publish it).

Another important document that is not always specifically requested is the "workflow diagram." This is a bunch of boxes and arrows explaining the process of an application and how it flows from one screen or process to the next. These are especially helpful if you have a unique signup process or if a user has to complete a number of steps before you permit access to a specific area or tool. Microsoft Viso is the tool of choice for developers and managers alike who use Windows operating systems. For a great Mac product, check out Omnigraffle (it can also help you with the wireframing process).

Finally, in any scoping process, there has to be a **deliverables statement**. This is something that you and your developer work on together to define exactly what you're going to get at the end of the process. This not only includes what code you get to keep or how you judge the "completeness" of the project, but also things like progress reports, weekly meetings, development milestones, and estimated completion dates. A lot of these items are ordinarily held over until the contract signing, but the sooner you tackle them, the better.

Kick-Off

"You will never see eye-to-eye if you never meet face-to-face." —Warren Buffett

Once you gather all your requirements together in a scope and you pick out an awesome development team, the ball starts to roll. When working with a professional team, this means starting with a kickoff meeting to discuss the way the project will run. During this meeting, cover a few essential bases to keep your expectations in check.

The Goals

What are you planning to achieve? It's best that everyone involved, from the project managers to the coders themselves, be aware of the reason for building this software. Take a few minutes to outline the pain point or the opportunity the project addresses. Doing so gives context and meaning to the team members. It's surprising how often engineers focus on solving a problem for the sake of solving it. Having an understanding of the "why" is extremely important. Programmers make hundreds of small decisions during a development project. Informed decisions strengthen your chances of success. Try to put ideas in terms of the end users—the actual people who will interact with the product.

The Team

Make sure you know who will be responsible for what. If you can shake the hand of each person, do so. A little bit of the human element goes a long way, even with programmers. I'd avoid hugs though; some people are weird about that.

Communications

Have a clear chain of command and get the contact details of everyone with whom you will need to be in contact. Project managers are there to protect the project owner from the programmers and vice versa. Software engineers are not divas or delicate flowers, but teams have their own type of internal logic. Rely on your project manager. They know how to communicate with their developers.

Setting up a regular check-in call is highly recommended. This doesn't have to be any longer than a planned fifteen minutes every two weeks. Communication becomes more frequent closer to delivery. This is because your involvement increases the more there is to see and touch. The early stages of a software project involve things like setting up the system's architecture and designing the database—steps that are tricky to evaluate or assist with.

Tools

The tools part of the conversation is where the nerds get excited. Make sure you know exactly which communication tools you'll use. What type of system does your PM use for communication purposes? How will you notify the team of bugs you find? Most development teams have a favorite ticketing or bug tracking system. Schedule some time during the kickoff for a little training on the system of choice. Be very wary of anyone who suggests that "email is probably the best way to go." Email has a nasty way of going missing when something goes wrong! Ticking systems allow for tracking of tasks, bugs, and communications.

The Project Plan

This is where you see the organization level of a development team. At a minimum, ask for a clearly defined set of milestones, including their expected dates and what to expect for each milestone. This might be outside the scope of the kickoff, but make it clear that you expect it soon. Depending on the team's project management methodology, this may include shiny things like Gantt charts, project outlines, or review timetables.

Metrics

Whether the goal of your system is to increase your customer base, reduce call times, improve sales, or even to improve the experience of your guests, you can measure progress in some way. The kickoff meeting is a great time to discuss how the team will incorporate the analytics you need to measure the success or the return on investment. Not every project has easy-to-design metrics (like "number of user sign ups per day"), but knowing the possibilities helps drive the developer's intentions.

UX and UI

"A designer knows he has achieved perfection not when there is nothing left to add, but when there is nothing left to take away." —Antoine de Saint-Exupery

User experience design is one of the most important exercises in the entire development process. This entails brainstorming and planning how the user will interact with your application and what that will look like in detail. As opposed to the scoping of the application, UX design maps out and represents the visual flow of an application. This is then given to your programmers as a blueprint to follow.

Application Flow

Let's design an imaginary application, shall we? I'm a big whiskey fan, so we're going to design a whiskey rating app. If that doesn't appeal to you, substitute "whiskey" for "weasel." This will take your mind off the problem, and most likely, evoke some interesting imagery. Anyway. . . .

We'll start with a very basic outline of the movement from one screen to the next. I'm going to illustrate this below with a flow diagram.

As you can see, this app might not have a lot going for it, but at a quick glance, a programmer can tell a lot about its intended function. Putting myself in the programmer's shoes, this flow diagram now becomes my to-do list. I know I have to design a login page, a sign-up page, a profile page, and so on. Unless it's a pretty basic website, I need to design each one of these pages. That might sound excessive, but trust me; without a thorough blueprint, you're going to leave a lot up to the interpretation of the individual developer. While your programmer might shine when it comes to inventing each item required for each screen, having an agreed-upon blueprint reduces tensions down the line.

Wireframing

Now that you know which screens you need to build out, the next step in the process is getting each one sketched out. By that, I mean actually drawing each of the screens which a user will interact with. The term wireframing has already come up a few times in this book. Put simply, the word originates from the fact that most of the time, these sketches look like pieces of wire outlining the most important parts.

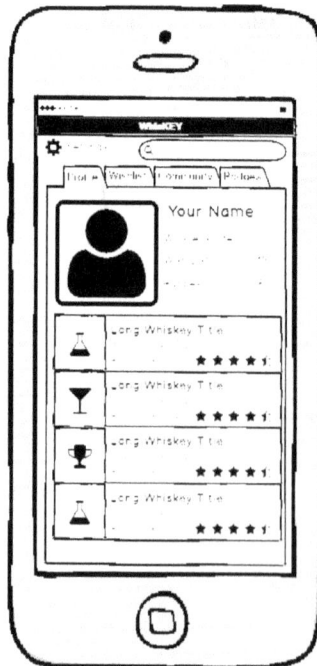

You can make wireframes with pretty much anything. A good set of pencil sketches on white foolscap is a great place to start. Microsoft Powerpoint is also a popular albeit imperfect tool for creating wireframes. There are a number of really good and easy to use software applications out there to make the process a little easier. These programs include libraries of common elements you would see on a website, mobile app, or desktop application. Things like scroll bars, buttons, forms, and so on, are all pretty much drag-and-drop. Currently, my favorite wireframing tool is Balsamiq (http://balsamiq.com). It's easy to use and as close to a standard in the industry as you can get.

Your developers should be the ones making the wireframes, then consulting you for feedback. If you want to take an active role in the wireframing process, feel free! It's a lot of fun (at least until you have to start designing the help screens). I would advise having your developer's UX team do most of the heavy lifting, even at the expense of cutting costs.

This is a three-fold strategic decision:

1. In building the "wires," your developers show you how they think and prove they know what they're doing.

2. Hopefully, you're paying the developers because they're experts. Getting into the actual flow of a system can be quite daunting, and unless you've got a lot of experience creating these types of documents, they're not as easy as you might think.

3. While your developers build out the wireframes, they might come up with creative ideas that never occurred to you during the initial planning. Make them earn their money!

When building a wireframe, keep in mind that you're trying to get the experience of the application down. You are not trying to do the graphical design of the entire product. Wireframes go to both the developers (so they can understand what they have to build) and to the designers (so they can see what elements they have to design). Wireframes should not look like the end product, just the idea of the application. I prefer wireframes that look like sketches on the back of a napkin because they don't give the designers too many preconceived ideas as to what you want your final product to look like, and you can maximize their creativity as a result. More on this in the next section.

In my whiskey rating application wireframe, I think I've done a respectable job of showing what needs to happen on this profile screen. There's a search bar, a profile picture, some tabs as my navigation system, and so on. I'm not 100 percent happy with it, but it's good enough to take to a designer. A good designer takes the elements that I wireframed and builds an attractive design, or skin, around them.

UI

"If the user is having a problem, it's our problem." —Steve Jobs

When someone uses the term UI or "user interface," they mean the graphical elements of a design. That is, if they know what they're talking about. It is easy to get confused when people jumble UI and UX in a sentence, which happens a lot. UX refers to blueprints and wireframes, and UI is the paint on top of the sketches that makes them look great.

User interface design is an entirely separate process from UX design. In our company, they are handled by different people with different expertise. Our UX designers are software experts engaged in requirements gathering and long client meetings. Frequently, we first sketch out UX as a collaborative process on a whiteboard in a discovery meeting. Then a UX designer steps in to polish, document, and organize the information.

Once there are enough wireframes for the application to start taking shape, the graphic designer, or UI designer, comes to the table. This process can be tricky due to the subjective nature of design. As a result, you might want to consider hiring one company to handle your UX and programming, and a separate company to handle the design. I've worked with many larger organizations who split all three functional processes between three separate companies. This is only practical with very large projects with equally sized budgets.

The graphic design of an application ranges from extremely basic to highly complex. In many situations, if the UX is simple enough, you won't need a designer at all. Modern programming frameworks include carefully thought out visual elements. Developers can use these frontend frameworks while they code, and you will generally get a fairly visually pleasing end product. A few of the most popular frontend frameworks are:

- Bootstrap (http://getbootstrap.com/)—The framework from Twitter that seems to run half the world, these days.

- Foundation (http://foundation.zurb.com/)—Only slightly less popular but still very much a go-to.

- Material Design Lite (https://getmdl.io)—Google's new design language-centric framework.

The other low-cost option is to go with a pre-built template. You can find templates available for almost every imaginable use case online. These will not be unique, but they get the job done. You can find these templates all over the web, ranging in price from a few dollars to thousands. My two favorite resources for templates are:

- ThemeForest (http://themeforest.net/)—My first choice for templates and quick start kits.

- TemplateMonster (http://www.templatemonster.com)—By no means new to the competition, TM offers a number of great resources.

Before I get stalked and murdered by a deranged graphic designer for my statements, I am only recommending these techniques if the application is not client-facing. If the application is only available to an administrative user performing some basic task, you probably don't need to spend money on a full design phase. If the public will view your application in any way, I highly recommend bringing in an expert designer.

Mockups, Comps, Wireframes

The first step in a UX/UI design is "wireframes," or "wires" for short. These are the sketches discussed in the previous section, sometimes built using a special wireframing tool like Balsamiq.

Mockups, sometimes shortened to "mocks" are the actual designs themselves. These happen in a graphic design application like Adobe's Photoshop or Illustrator. They look extremely close to the final product.

On occasion, you'll hear the term "comps," short for comprehensive designs. This term technically means that the design is one more stage further along than mockups and looks exactly like the final design. However, mockups and comps are often used interchangeably by designers.

Prototyping

"I have not failed. I've just found 10,000 ways that won't work." —Thomas Edison

Prototyping is the final, and often overlooked, phase in the planning process. Building a working version of something before you produce it is common in the manufacturing world, but it doesn't always make sense in software development. Instead, when we talk about prototyping in programming, we're talking about clickable wireframes or designs.

With prototyping, you see your cocktail napkin sketches or app design on a screen, and you can then interact with it in a very basic manner. I know this may seem like an odd and possibly redundant step, but it is incredibly valuable. While wireframes are great, you can't really test them on other people without explaining every step of the process. Designs look pretty on paper, but are they useful in the real world? This is a problem for larger, more complicated systems. A button labeled "Save" might make perfect sense from a design standpoint, but at a functional level, a user might expect it to not only save their work but also close the screen—a subtle but important difference. Doing simple user testing with coworkers, potential clients, friends, family, or just random retirees you meet at the airport terminal is a telling method of finding out what users think of your application.

If you want to tackle this at the wireframing stage, most modern wireframing tools incorporate the ability to make these kinds of prototypes. Balsamiq (which I mentioned earlier) has the ability to do this with some basic linking. It will even generate clickable PDF files that you can email to unsuspecting guinea pigs for free testing. There are a number of other great tools out there to make more fleshed-out, interactive wireframes, like Generator (http://designmodo. com/generator/) and InVision (http://www.invisionapp.com/), but these are a little more involved than you need to get. Another great tool we use is Marvel (https://marvelapp.com). Marvel is an app that allows you to take photos of pen and paper sketches and then link areas of them together to make an interactive prototype. The other nice thing about Marvel is that if your designs are already done, you can use those instead of the wireframes.

The other common prototyping technique is to develop the frontend of the application you're going to build. This literally means coding the design part of the app all the way from concept to a working product. The goal is to create a basic version of the app or website that runs on the device intended for use in the final stage of the process. In fact, prototyping this way has the added benefit of doing some of the work needed in later phases. The programmer works with the designer in building each screen or "view" in a programming language rather than a graphics program. They link buttons and boxes together using a simple linking code, so that when you install the application or play with it in a browser, it will appear to be the real deal.

This approach has some pros and cons. On the pros side, this tactic forces you to look at your user experience in extreme detail. Building the frontend out as a prototype eliminates excuses for not getting it right from the get-go. You'll interact with the interface you will have in your final version—not an approximation—so you complete a lot of the frontend work in advance. Once you build out a working version of the frontend, you'll be able to show it to stakeholders and let them play with something that looks and functions like the finished product.

On the cons side, this approach is more time-consuming. Without a doubt, it's slower and more expensive. Once you build all these real-life screens, it becomes more difficult to make changes (at least compared to a set of wireframes anticipated to change often). Plus, a clickable wireframe or prototype carries a psychological air of being "just a blueprint." Playing with these "blueprints" makes users gloss over minor issues because they assume you will correct them in later stages. This frustrates developers and is the number one reason for scope creep. Finally, an issue arises with expectations. If you put a prototype that looks like the finished product into a manager or investor's hands (especially one who is not technically inclined), they will tend to mistakenly believe that the product must be pretty close to being finished. Explaining to them that this is only the tip of the iceberg in terms of completion may be tricky. In this scenario, you might get outbursts such as, "I saw the damn thing working five months ago, what is the holdup?"

QA

"Art is never finished, only abandoned." —Leonardo da Vinci

Quality assurance, or QA for short, is the process of testing your software to make sure it works as expected. Unfortunately, no software is perfect. I'm sure you've had software products built by Fortune 500 companies crash on you. If hundreds of developers, designers, and QA professionals can't make something perfect, it's likely that anything more complicated than a basic calculator will have problems. But don't despair! With a bit of time and effort, you can get an application to a point where any issues are close to unnoticeable.

Done separately, as part of the traditional waterfall project management cycle, QA is consistently the longest part of the development process. You now have a large, basically complete piece of software to test, and you're stuck in a cycle of grinding out requests to developers, waiting for fixes that might take weeks, and then repeating the process over and over. It's extremely frustrating to see a project that looks so complete but, once tested, reveals hundreds of bugs. You can mitigate your frustration with a more continuous QA cycle. Communication with your team becomes paramount, and testing smaller, more manageable chunks of the project helps overall quality and your mental health.

We're all familiar with **bugs**—a word that has been used to describe a fault with a process or a machine since the 1800s. Grace Hopper (a pioneer in computer engineering) cemented the term into computer engineering jargon in 1947 when her team discovered a moth in one of their computers relays. If you're ever in Washington, DC, you can see the actual moth taped into Hopper's log book at the Smithsonian's Museum of American History. I have a few cheesy photos of myself and the famous moth, but mainly to entertain other nerds. Nowadays, it's pretty rare for an actual insect to be the cause of a technical problem, although if you Google "insect in computer," you will find some pretty horrific stories.

Software bugs don't exist because programmers are lazy, stupid, or incompetent (although those can be reasons). The three main reasons that you'll find bugs in your software are:

1. **Complexity.** Large software projects can be so large and complex that it is almost impossible to anticipate the effects a change in one part of

the code will have on another. These changes manifest as bugs in parts
of a system that were once considered bug free. This means you will
often have to do "regression testing," which means testing everything
over from the beginning every now and again rather than just testing a
single part.

2. **External libraries.** Your developers will not write every part of your
    system. In fact, much of what modern developers do is assemble
    prewritten libraries of other people's codes into a single package. This
    cuts down development times by up to 90 percent. With that kind of
    time and cost savings, this is not going to change anytime soon. On
    the other hand, it becomes hard to tell how these various libraries are
    going to interact with each other, and if there is a bug in the library
    your developer is using, it becomes even harder to track it down.

3. **Requirements issues.** Sometimes, a bug is not a glitch. It can be a
    system that works the way the programmer intended it to, but not
    the way stakeholders wanted. This is due to miscommunications or
    poor planning. The software and business worlds move very quickly.
    Change requests come hard and fast. In environments where this is
    unavoidable, managers and developers have to realize that this will
    cause issues somewhere along the line, and they have processes in place
    to mitigate these risks. In newer projects, poor planning is the number
    one reason for these types of bugs.

Types of QA

Internal QA Teams

Many development companies or teams have a dedicated QA team or QA
person. Sometimes your project manager may double as a QA lead or the QA
team in its entirety. A development agency might add these QA teams as a
separate line item in a quote. This is because when developers give their project
managers estimates for the build of a system, they aren't taking into account the
back and forth of changes and bugs that inevitably arise in the development
cycle.

A good QA person is worth their weight in gold. They are most likely a technical
person, capable of doing more than checking on whether something works as

expected or not. They will accomplish tasks like:

- Check any APIs created by the developer, making sure they are private and not open to the public
- Create bug reports for developers that programmers can easily understand in their language
- Do some basic security checks and report potential vulnerabilities
- Draw on their experience to find problems they've seen before

Outsourced QA

In some situations, software teams might not be large enough to have their own QA staff, or they might be working on projects so large that no one person or even one team of people is enough. There are a number of QA services out there that developers will rely on to either augment their QA teams or to handle it completely. Services like Applause (https://www.applause.com/) often supply developers with teams of QA professionals who are experienced in finding bugs, and who offer insight into UI/UX and prioritizing issues. Another nice feature of these outsourced agencies is that they have access to hundreds of different devices, running any number of different operating systems, web browsers, and screen sizes. This means that they will have a higher success rate in finding issues that span across platforms. It's common for a web application to work perfectly in Chrome, then not work at all in Safari. Microsoft's Internet Explorer is especially feared among developers for its inconsistencies between versions and large discrepancies with other browsers.

You!

The first line of defense in any software project is you! You are the boss, subject matter expert, or simply another pair of eyes, and many developers rely on at least some level of QA from their stakeholders. This means that your developers will rope either you (or someone on a nontechnical team) into the QA process. This is not as bad as it sounds. The subject matter experts who helped during the planning phase are ultimately the QA team who will have the final say. Make sure you're happy with what you've paid for.

Hallway Testing

My favorite type of testing is the unsuspecting victim I find in the office, or preferably at a local house game of poker. "Hallway testing" refers to grabbing people walking by to test your application. My friends often vanish to refresh their drinks or simply fold and walk away when I pull out my cell phone and exclaim, "Hey, check this out."

Taking an application out into the real world is a must for any software testing. The results can be humbling due to the fact that you, your developers, and your QA team are all close to the project and may not see the forest for the trees. Watching a user interact with a piece of software for the first time is the most valuable type of testing, especially for the UX of the system. Watching a first-timer fumble their way through an interface lets you know exactly what to improve. Another nice plus is that it's usually free, although if you abuse it too much, it can cost you friendships.

QA Tracking

- Jira
- Trac
- BugZilla
- Redmine
- Trello

These tools are very boring but very beneficial (and fairly easy to operate). You or your staff can log in to the system your developer uses and create "tickets," which explain an issue that you are seeing. All these systems use a workflow process that typically moves a ticket from the person testing the application to a developer, and then back again. Each time the ticket moves from one person to the next, the status of the ticket changes, alerting the next person in the workflow what is going on. None of these systems are exactly the same, but they each have a status and workflow status similar to the following:

Tickets are either:

- Open (New.)
- In progress (Someone is working on them.)

- In QA ("We think we've got the issue fixed and are waiting on you to confirm.")
- Fixed (All done!)

After the creation of a ticket comes categorization. These categories help project managers and developers know if something is wrong or if a ticket is some other type of communication. Because ticketing systems are regularly used in both the development process and the QA process, this helps separate items. Some development teams have support licenses that are separate from their development billing, too. This means that the category of a ticket you file might be billable or included in your support agreement.

New ticket classifications include:

- Bugs (Something is wrong!)
- Tasks (This isn't a bug, and it's not a new feature. These are things like text changes for a page or a new image.)
- Improvements (A tweak to an existing feature.)
- New Features (Something that hasn't existed before, but should. This may or may not require a change order and extra billing.)

Finally, tickets have a priority. Priorities have a few flavors as well:

- Blocker (The system isn't working, or you can't continue testing because of this issue.)
- Critical (The application can't go live without this.)
- Major (Most issues are major—this just means it's important.)
- Minor (A nice-to-have or a small fix/add.)
- Trivial (Either something very easy to do or something that isn't required for the next version of your software.)

Once the QA person (or yourself) classifies the issue, it will assign it to a developer or a project manager. In larger projects, the PM will triage the ticket (work out how important it is and put it in the work queue based on priority) and then assign it to the lead developer or a specific developer on the project. When the developer completes the task, they will change the ticket's status to

"Awaiting QA," and the PM will assign it back to you or the QA person to test it out.

QA can be a fairly tedious and sometimes frustrating task. There is a delicate balance between getting something polished enough to ship out and trying to reach perfection.

Talking to Developers: Tips and Tricks

"Most good programmers do programming not because they expect to get paid or get adulation by the public, but because it is fun to program." —Linus Torvalds

There are a few things to keep in mind during the development process that don't really fit into any of the previous chapters. So this is a small smorgasbord of tips and tricks to guide your conversations with your developers.

Asking Questions

Trust that the team you picked are experts, but ask lots of questions. Communication is key in any complex build process, but doubly so when neither party has a deep understanding of the product. Unless you've found a developer who specializes in mobile apps for salespeople in the aluminum recycling industry, it's unlikely that your team will be experts in what you do. There is only so much that you can tackle in the planning phase, and both sides generally come out thinking that they know exactly what to expect. Unfortunately, this is not normally the case. All systems contain unexpected complexity. The best way to deal with this is to ask lots and lots of questions about the process. You don't have to be a Nagging Nelly, but get involved. Try to participate in a telephone meeting at least once every two weeks at first, and increase that to once a week when there is something to see and touch. Try and make a face-to-face meeting once a month if possible. Very little substitutes for an in-person meeting.

Documentation

Developers live and die by documentation. When a change to the plan arises, or you require clarification, write it down in detail. This is not just to protect your investment, but it increases the trust between you and your team and lets them review the history of changes. A good development team gives you access to some sort of shared document management system or wiki. Learn to use it. It will give them one less thing to complain about, and that shields you from excuses down the line. Find a tool that you're comfortable with for annotating documents (Adobe Acrobat is a great one) and an application for making your own mockups or wireframes. This will help you communicate to your team what you want to be accomplished.

Giving Feedback

Feedback is critical in any project. I wish developers could read minds. But if we could, we'd probably be in the stock market rather than in software development!

Feedback requires a balance between what you, the client, want and how the software developers think it should work. The best solution is somewhere in the middle. That doesn't mean your developer won't always let you win in the end, but you want to hire an expert, not a yes-man. Should you hear a developer eventually sigh, and then say, "Okay" to everything, have a look at http://theoatmeal.com/comics/design_hell and realize that you're paying good money for a group of developers who have done this before.

During every phase of development, you should get your team to do a show and tell. This means they present you with scopes, wireframes, small demos, and more and more completed versions of the software to test and provide feedback on. That leads us to the question of how to actually provide feedback. This isn't as intuitive as it may seem. Feedback is something that we all give all the time, but when it's about something specific, like how an app should function, there are some general guidelines to keep in mind.

Be descriptive, not prescriptive

"The buttons should be a darker color." This is prescriptive feedback. It's hard to quantify and can hurt a designer's feelings (artists are sensitive souls).

"I think the button is a little too light for the text on top of it. Our audience is typically older, and low contrast is harder on their eyes." This is descriptive feedback. It lets your team know the thinking behind an issue and gives them the space to play with a fix that fits with new information while still deriving value from their knowledge base.

Don't micromanage, but don't macromanage, either

Oh no! Middle path nonsense alert! This is a hard one, but it's important. Micromanaging is totally understandable when there are clear brand guidelines or legal documents that have to be exact. But on the opposite end of the spectrum are clients who say, "Just do whatever you feel is right." In theory, that would be lovely. But you have an opinion, and your developer would love to hear it! Learn to use something like Powerpoint or Balsamiq or anything you feel comfortable with, which you can use to add elements on top of a screenshot. Then you can add notes in the margin and send them to your team.

Use past tense, not present tense

It's hard to believe, but this makes a whole lot of difference in the perception of your feedback. Speaking in the past tense helps avoid conflicts and will focus the discussion on why something made someone feel a certain way.

For example:

- **Makes a developer feel bad**: "It isn't clear what to do after you tap the first button."
- **Makes a developer feel good**: "It wasn't clear to me what I should do after I tap the first button."

When you hire a development team, you're paying for expertise, not hiring a monkey to say, "Yes, sir/ma'am," to every request. (Though I concede that if monkeys could talk, they might not be that polite.) You should trust your team but give constant feedback. At the end of the day, you are the subject matter expert, and they are the development experts.

Finally, learn to accept feedback from your team. There is only one way to do this: "Thank you. That's good feedback. Let me think about it." Any feedback is probably good feedback. If the feedback really is helpful, great! If it seems silly, it might mean you didn't get the point of the application across to the team. If the feedback is about out-of-scope future phases, just make a note of it and think about it later. If the feedback is off topic, keep in mind that it might be a symptom of something else going on. There is no such thing as bad feedback if you take this approach.

Software in Real Life

Skype demos and videos are really important ways for your programmers to show you progress, but seeing something in the real world is extremely important. You should be able to take a version of whatever your team is working on home with you. This might be an early-stage app on your personal phone or a web link to a beta product. If you can, spend a little time in the same room with your team while they watch you or someone you brought with you play with the application. This will humanize the users and allow your team to see where UX issues cause problems.

PART III
LAUNCHING AND BEYOND

CHAPTER 10
DEPLOYING APPLICATIONS AND DISTRIBUTION

"How much for that software in the window?"

I haven't walked into Staples and picked a box of software off the shelf in quite some time. In fact, "a box of software" felt weird as soon as I typed it. As I mentioned earlier, my laptop doesn't even have a disk drive, so buying anything that came in a physical box would be unlikely to help me in any way. That doesn't mean that software isn't still sold in boxes, but let's just say I wouldn't invest in any company still making physical media.

Most software changes hands over the Internet. Websites offer downloads, and the Apple and Android app stores sell billions of apps over cell tower connections. Distribution is easier than ever, resulting in a very competitive marketplace.

In the following two sections, we'll look at the two most common ways to get your software into the hands of your eager users and take over the world.

The Web

Websites, email servers, online games, and every mildly interesting thing a cat has ever done lives on the Internet. But what does that mean, really? How is something "on the Internet?" Engineers refer to the process of getting your software and databases onto the web as **deployment**.

If you think back to the chapter on "The Stack," there are two steps to putting something up online. First, you need a computer permanently connected to the Internet. We could go into detail about how all that works, but all you really need for this is a static IP address (a network address you purchase from your Internet provider that doesn't change). This permanently connected computer is your server. Second, you need a piece of software on that server to listen and respond to requests. To make it extra confusing, this piece of software is also referred to as a "server."

You can deploy your applications in a number of ways. In the old days of web development, you connected to your server, set up its databases, and then copied-and-pasted your code over to the physical machine. It was easy to make mistakes with this process.

A programmer moving code from one system to another eventually makes a mistake. They might copy something into the wrong place or overwrite the wrong file.

Things are (or least should be) a lot better these days. We maintain and manage code in repositories, which track all the changes each time a programmer touches a file.

App Stores

"Last Wednesday, I stupidly dropped my iPhone in the bath, and my life has sort of spiraled almost out of control." —Patrick Stewart

If you're going to publish a mobile app, the process is a little different from publishing on the web. Here, I'll focus only on the Apple App Store and the Google Play Store because they're the two giants. If you're going to publish a Windows Phone app or a Blackberry app, the process will be similar.

Google Play

Android apps come bundled in files called APKs (Android Package Kit). Android makes it easy to test apps by allowing you to install an APK file on your phone right from your email or a download link. You will need to make a few changes in your phone's settings to allow this. In your phone's settings menu, navigate to "Security" and check the boxes next to "Unknown Sources" and "Verify Apps." These two settings allow you to install apps from locations other than the Play Store without throwing all sorts of errors and warnings at you. You can always turn these off when you're done testing for security. Because Android makes it so easy to install apps, you could accidentally install something nasty from the Internet without these settings turned off.

After all your testing and QA, you'll want to get your app into the Google Play Store. This will allow the world to get their hands on your new shiny app!

Signing up for Google Play is extremely easy. You will need to visit https://play.google.com/apps/publish/ and log in. All you need is a Google account (Gmail, Google Plus, and so on). You need to agree to some terms and conditions and then pay a $25 registration fee. Finally, you need to fill out some information about your business and agree to some more terms and conditions.

After that's done, you shouldn't have much more to do. The main thing is to go to the "Settings" option on the left-hand sidebar. Choose "User Accounts & Rights" and invite your developer to the account. They can add your files and screenshots needed for the Play Store. In the beginning, give them the role of "Release Manager."

There isn't much of a Google Play Store review process. They run some automated tests on your app to make sure it isn't completely evil. Other than that, you are free to do what you want. You can even upload all the adult content or toilet-humor apps you want.

Apple App Store

Ah, Daddy Apple. Apple is not like Google. Apple tightly controls everything about their ecosystem, from the content of your apps to the way they function. The rules for what you can and cannot submit are purposefully vague. This is likely due to the fact that Apple wants to reserve the right to kick you out of their "walled garden" for any number of reasons. Installing an app on your iPhone or iPad is equally difficult for the non-tech-savvy.

An Apple Developer account must "sign" every iOS app. In essence, this is a security certificate that Apple gives your developers, unique to each app. You can sign apps for testing purposes or for production (when your app finally goes into the store). Your developer will almost certainly have their own Apple Developer account to get this done for you. Still, it's a good idea to have your own Apple Developer account. You will need one if you plan to charge for your app or sell in-app purchases.

Even if you plan on giving your app away for free, you should have your own account, as it gives you control of your property if you ever need to switch developers for whatever reason. Companies go out of business or get acquired. Relationships may sour. Or your programmer might get arrested for jumping into a gorilla cage at the zoo. Either way, it's better to be safe than sorry.

An Apple Developer account costs $99 a year, but that's not all! You'll need a Dun & Bradstreet number (D-U-N-S), which you can register for at https://www.dandb.com—this can take up to two weeks. In order to get your D-U-N-S number, you'll need to have a legal business entity, LLC, S-Corp, C-Corp, and so on. Once you've got all that covered, you can go to https://developer.apple.com and sign up for your developer account. Oh, wait! That registration process can take a few days too. All in all, you should give yourself a good month to prep your Apple account.

Apple bundles apps as IPA files. Unfortunately, this has nothing to do with beer, although they do drive some to drink. IPA stands for "iOS App Store Package." You can't install an IPA file from the web or an email link. There are two main ways to test an IPA file on your Apple device.

1. The traditional way to install an IPA file is to send your developer your UDID number. This is a Unique Device Identifier, which is only for your particular iPhone or iPad. Each device has a different number. You can find it only when your device is connected to your computer and you have iTunes open. There is a nifty website that will guide you through this at http://whatsmyudid.com. Once you receive your UDID, you send that off to your developer. They add this to their Apple Developer account and can send you an IPA file that only runs on the registered device. Install your IPA through iTunes using this method. You have to open the IPA file with iTunes and then sync your device. Then, and only then, will you see your flashy new icon on your iPhone or iPad. At this point, you most likely need that drink, but shudder if anyone offers you an Indian Pale Ale!

2. There is, thank heavens, another way. In 2014, Apple bought a company called TestFlight, which allows developers to send regular folk a simple install link. All you have to do is download the TestFlight app and wait for an email from your developer. Click on the link to your phone's email application and TestFlight will open up, allowing you to install your new application without so much as a complaint. Make sure your developer is familiar with TestFlight—it will make your life significantly less painful.

When the dust settles and your app is built, tested, and ready to go, you'll need to submit it to the App Store. Your developer will likely need to do this for you, but for once, that's easy! All you need to do is log in to your Apple Developer account, click "People" on the left-hand sidebar, and invite your developer as an "Admin." They should be able to do the rest.

CHAPTER 11
MONETIZATION

For some of us, the creation of a software is a pure art form. We do it for the fun, the challenge, the fame, or the good of mankind. The rest of us do it for the money.

Unfortunately, the concept of "if you build it, they will come" is a myth. (I'm sure it's worked in one or two situations, but I've personally never seen it happen). The hardest parts of business are always the same: marketing, branding, sales, customer service, and human relationships. It doesn't matter if you sell skateboards in a brick-and-mortar store or online; the hard work begins after you build everything. The idea that a software company is easier to build than any other business is an absolute fantasy. Now, this is not a book about sales and marketing, but there are enough small differences in the way one sells software that these concepts need to be addressed.

In this chapter, we'll look at some of the ways you can make money with your application. We'll look at the app store models, as well as some common methods to generate income using the web.

Making Money in the App Stores

If you're building a mobile app, your revenue mainly comes from your publisher: Apple or Google. There are four main strategies for making money in the app store:

- Selling your app directly
- In-app purchases
- Traditional e-commerce
- Advertising

Selling Directly

If you've ever bought an app from the Apple App Store or the Google Play Store, it's likely you paid $0.99. That's because it's the most common price for apps out there. As you can imagine, it's fairly tough to make a fortune selling anything for a dollar. What's more is that Apple and Google take 30 percent of everything that goes through their stores. That's right, ladies and gentlemen, 30 percent for everything. It's even tougher to get rich selling anything for $0.70 a pop.

In a hypothetical fairyland of entrepreneurial bliss, your first new app is a smashing success. You wake up to find 100,000 downloads. Apple features your app on the front page of the App Store. Google calls you up to say they want to help you promote your app, for free! Let's run some conservative numbers in fairyland here.

Say your app cost $40,000 to develop, and you put $10k into marketing it. Your 100,000 downloads generate $69,300 in revenue to you after Apple's take. You net $19,300, which is really good. If the app uses any server technology (which it likely does), your burn rate is roughly $1,000 a month in server costs. If your 100,000 purchases are over the course of a year, you are now left with a profit of $7,300 . . . which doesn't look like a great return on investment for a $62,000 outlay. These are generous numbers.

Not to terrify you completely, but users also expect fairly constant updates, new content, and added shininess. That means more development costs.

You either need to shoot for 1,000,000 downloads (tough) or find some other way to make money.

Pricing

If your goal is to create a little more perceived value, a paid app is the way to go. The obvious way to make more cash is to increase the price. A few rules of thumb:

- Business applications command the highest margins of $10 to $50 per download.

- Social apps range between $5 and $10 a download, but anything over the $5.99 range needs to have some serious perceived value.

- Developer tools (apps for programmers) can produce a fair amount more—$5 to $50.

- Games are a bit more difficult to quantify. Puzzle games are always the most popular but generally are premium (we'll get to that in a bit) or inexpensive, with in-app purchases (stay tuned). Big 3D games from huge game studios are closer to the price of business apps.

The key here is to experiment. You probably want to launch your app at a lower price, then raise the price as you gain users. If you can hit 200 to 500 downloads a day, that's usually a good sign. Once you hit that 100,000 mark, you can try lowering your price again or doing a promotional deal to pick up a higher market share.

In-App Purchases

There are two ways of selling to your users once they begin using your app. The first is "in-app purchases" or IAP.

This payment method is native to your device. When you hit the "buy" button in the app, the payment flow is the operating system's standard payment process. You can't customize the way it looks or feels. Apple and Google both offer in-app purchase tools for developers, and the concept is almost exactly the same on both platforms. Each time a user hits the "buy" button, the publisher bills the user's credit card associated with their Apple iTunes or Google Play account.

The developer accounts get credited the full amount, minus 30 percent. The app stores make between three and four billion dollars a year on that 30 percent, which sounds like a lot, but it accounts for less than 2 percent of revenue for each company. That gives you an idea of how much money Apple and Google make.

In-app purchases are normally used to enhance some feature of the application itself. This could be an ongoing subscription to a service (like a dating app) or enhanced features (like LinkedIn). Games are the big winner in the in-app purchase category. Games use IAPs for allowing players to unlock new characters, buy extra lives, or open up new levels.

There are a few issues with in-app purchases, though. Keep these in mind when designing your app. As mentioned previously in this book, you can't sell "real" items with IAP. Virtual items (like new game characters), subscriptions, or digital downloads are fine. But you couldn't sell physical, real-world items like groceries or books. Why? Apple and Google provide developers with a digital delivery platform, and you with a payment method. This means that they are responsible for the end-to-end architecture that allows these purchases to occur. If a product or service strays outside of the ecosystem, the app stores stay out of it. This is why you have to put in your credit card for the Uber or Lyft apps.

If you want to see how tricky this becomes, take a look at the Apple vs. Amazon struggle from a few years back. Amazon's audiobook sister company, Audible, is the largest audiobook store in the world. There is a great Audible app that I use every day on my commute. I can use the app to search through their catalog and even listen to audio samples of the recordings. I cannot buy anything using the app. Why not? Well, Google and Apple classify audiobooks as digital goods and, as a result, take a 30 percent cut of each sale. Don't get me wrong—under the app store rules, Amazon could happily sell their audiobooks to me inside of the app using IAPs. But Amazon understandably doesn't want to take a 30 percent bath on every book they sell. So if I want to buy something I've added to my wish list, I have to visit the Audible website and punch in my credit card number. Once I've checked out, the book magically appears in my Audible app library.

A lot of apps on the market fall into the freemium category. This is a combination of free and premium. The basic concept here is that I give away the app for free and then try to up-sell the users with in-app purchases once they've seen the benefit of the app. A typical model is a free app with a bunch of annoying ads on the bottom of the screen. I can buy an IAP to remove the ads.

Freemium apps generate the least overall profit compared to the other revenue models out there. They provide a way to generate revenue from a diehard fan base while at the same time providing a risk-free way for users to try out a new application. Dropbox (https://www.dropbox.com) is a great example of freemium done right. Dropbox offers cloud storage starting at 2GB of space. That's probably a lot of space for most users and good enough for their needs. In fact, some sources report that less than 4 percent of Dropbox users are paying customers! Some estimates put their revenues at somewhere around $400 million a year. That doesn't mean Dropbox is profitable yet, but that's a story for another day.

Traditional E-Commerce

Back to our Uber or Lyft example. What if you want to sell car rides or physical books like Amazon, or even groceries like Instacart? Well, we know we can't use in-app purchases, so the obvious answer is "insert your credit card here."

This is your typical e-commerce play. Whether your app sells t-shirts or houses, it doesn't matter; these apps must use some kind of outside payment gateway like PayPal, Stripe, or Square. The only real reason to do this in a mobile app is for the convenience, rather than any real technological hurdle. Your user simply enters their credit card or bank information into the app the same way you would do it on the web. It might sound harder to do with an app, but it's fairly straightforward. Usually, the reasoning to go with an app here is part of a mobile-first strategy.

In-App Advertising

Ah, the oldest media moneymaker. Advertising. Also, possibly the least favorite side effect of any business that needs to make money.

Ads are everywhere these days. Google is, in many ways, an advertisement delivery platform, with 90 percent of their roughly $80 billion annual revenue comes from ads. All those fancy self-driving cars, fiber services, and venture funds are side projects funded by ads, so it's understandable that ads infiltrated the mobile space so significantly. However, advertising inside of apps accounts for only 15 percent of all app revenue. That's compared to in-app purchases and paid apps splitting the remaining 85 percent. Ads are a tough way to make money.

The conversion rate (or click through rate) of mobile apps is pretty good compared to the web. This might be due to the fact that smaller screens are more prone to users accidentally clicking on ads, but clicks are clicks. If we run the numbers on our free app example (with ads as the only revenue source and 100,000 downloads) our total revenues are more likely to be around $10k, much worse than selling the app outright, which earned us $19.3k.

You can hedge your bets here by offering an in-app purchase of a dollar or so to remove the ads (a common strategy). If 15 percent of your users convert, you're raking in another $10k or so after the App Store's take.

Taking a Cut

Companies who facilitate a transaction and then take a cut strongly believe in the old saying, "The secret to success in business is being a good middleman." This is quickly becoming a popular business model with the explosion in crowdsourcing and marketplaces. Essentially, you offer an environment for two or more people to offer each other a product or service, and you get in on the action when money changes hands.

Nerd Out

Crowdsourcing: Getting other people to do stuff for you. Lyft and TaskRabbit are two great examples. These are businesses that connect part-time workers to buyers to give them a ride or pick up their shopping. The company providing the platform to connect the buyer and the worker takes a cut of the transaction.

Marketplaces: Apps or websites where two or more people do business. Here, the people doing the buying and selling have nothing to do with the operation of the marketplace. The marketplace owners get a cut of each transaction. The classic example is eBay's auction site, but you can also consider traditional sales systems like Etsy and Shopify marketplaces.

Let's look at a hypothetical example: You have built an online Beanie Baby marketplace that allows people to sell their old would-be collectibles to anyone in the world. Collectors can use your app to take photos of their previously precious stuffed animals, set a price, and offer them up for sale on eBeanieBaby.com. When a guest browsing the site finally spots that orange polar bear they've been searching for, they hit the buy button and put in their credit card information. The overpriced chew toy is on its way! But what happens behind the scenes?

To understand how everyone gets paid, we need to look at the exciting world of "chained payments." Chained payments are the systems by which one payment gets split up into smaller payments and transferred to the correct bank accounts. There are a number of ways to handle these, but it can get quite tricky.

The most common way to process these types of transactions was to use a traditional merchant account from a bank and pay a credit card gateway to handle the transaction. When your Beanie Baby buyer entered their credit card, their money would go into one main bank account. EBeanieBaby. com would have to maintain records of the transaction, and then, once a day (or some other time period), cut a check or an automatic transfer to the buyer, minus the broker fee (eBeanieBaby's cut), credit card charges, and any transfer of check costs. If you're thinking, "That sounds so complicated and difficult that lots of things could go wrong," you're right. The other problem with this method is the bank starts to eat into either your profit or your sellers' profits, and nobody likes bank fees.

In recent years, payment systems grew much more sophisticated. Systems like Stripe and PayPal offer developers some tools to handle chained payments on a platform basis. In this scenario, your payment processor handles all the steps along the way. Normally, all three parties (you, the buyer, and the seller) need to have accounts with the payment processor, which takes the least amount of development time.

This means that you'll have PayPal or Stripe logos on your beautiful, professional Beanie Baby site, and these systems become directly involved in things like storing credit cards and handling refunds, payment disputes, account balances, and so on. When dealing with any of the above-mentioned features, your users interact with a third party and may have even left your site or app in order to do so.

Some people might be okay with that, but organizations often value the elegant over the practical. In using these out-of-the-box chained payment solutions, your developers can still hide all the stuff handled by the processor. But that takes work. Developers must write each of these systems by hand and then connect them to the payment processor through their APIs. This is not a large technical hurdle, but it adds time and cost to your project.

Software as a Service

Software as a service (SaaS) is the idea that a software application is centrally located and you typically pay a subscription fee to access it. That simple definition covers close to a $30 billion market. Online subscription systems are rapidly replacing most traditional software systems. There are a few reasons for this.

- Access to high-speed Internet connections is increasing.

- The power of the humble web browser has improved.

- Businesses like subscription models. Charging $200 for a piece of software is nice, especially if you can charge another $200 in three or four years for an upgrade. But charging $30 per month for the same software, and never requiring upgrades, nets businesses not only more cash but predictable, consistent cash flow.

- Users like subscription models (sometimes). For business clients, moving software from capital expenditures (larger upfront costs) to operational expenditures (ongoing, cost-of-doing-business expenses) can help stretch lean balance sheets. On the consumer side, it can be easier to express costs in terms of "only five cups of coffee a month" rather than "the price of a new TV."

Everyone loves SaaS! But there are some real business drivers that make it an extremely attractive model on top of just the pay-as-you-go pricing for your customers.

- The cash flow implications are huge. The math is simple, [subscription fees] x [number of users] = cash flow! Predictable cash flow is critical to all businesses but especially startups or small companies.

- Lower acquisition costs are a massive benefit. As already mentioned, it is much easier to convince a user that $19 a month is a no-brainer rather than $400 for a big piece of software. The psychology is a simple risk calculation on the user's side. While the buyer will know that they can cancel at any time, the inertia of actually deleting an account makes SaaS super sticky. When was the last time you switched cell phone providers? You know there is probably something better or cheaper out there, but the pain associated with switching keeps you "loyal."

- Tiered pricing allows businesses to charge for higher value, or more specialized features, to empower users without alienating the larger addressable market.

- Upgrades become easy to manage for users as well as the business behind the software. Rather than upgrading some users and supporting legacy clients, everyone gets the rolled-out changes at the same time.

Of course, there are some downsides to SaaS models. Nothing is perfect, except a good Islay single malt by the fire on a cold night.

- Web infrastructure is not infallible. Servers crash, outages occur, and when they do, all your customers start calling with complaints at the same time.

- Security becomes a serious issue. With every user's data stored on a centralized infrastructure, a breach can be catastrophic.

- Data mobility and integrations become tricky. If your niche SaaS system doesn't allow users to export their data into some usable format, or if it doesn't play nice with other SaaS systems, your customers can get nervous. What if you go out of business? What happens to all that data?

These concerns are all valid but are all addressable with an effective design, good data mobility policies, and expert infrastructure people. Just keep in mind that without covering these bases, you're opening your business up to some fairly serious risks.

Freemium SaaS

The freemium model became wildly popular in the last few years. The concept is simple: Give a basic version of your product away for free, and then charge your diehard base for additional features. Spotify is a great example of a successful freemium model. The music streaming service is free for anyone to use, but users have to listen to an ad every few songs. The people who are more likely to listen to Spotify all day long are the same people prepared to pay to eliminate ads. Everyone wins.

Freemium strategies hit that balance between user acquisition and revenue. By reducing the cost for potential users to basically zero, companies can quickly gain traction, prove marketability to investors, and build loyal user bases. This is the Internet age's version of shipping you a free demo disk. Even better, the traditional sales-driven marketing strategy of having to walk potential customers through a product demo goes away. Users who aren't paying anything are more prepared to work it out on their own. Dropbox, Evernote, Box, Pandora, and Spotify are all great examples of this.

The trick with freemium models? Balancing the cost of your free users against the profits of your paying ones. Servers, bandwidth, and support staff cost real money, so you can't have too many freeloaders. Free users often are great advocates for a new product or service, but they mainly attract other free users. This can cause a death spiral of costs. Free services also don't carry as much psychological value as pay-per-play alternatives. (If it's so great, why is it free?)

If you choose a freemium model, make sure you are positive that:

- Your service is ten times better than the competition

- The value of upgrading is significantly more appealing than alternatives

- You can survive on a low conversion rate (Dropbox converts only 3 to 4 percent of their free users to paying ones)

Advertising

In December of 2016, authorities busted a Russian hacking group for a massive web advertising scam. The idea was simple but executed on such a huge scale that they made between three and five million dollars a day!

The group registered around 6,000 domain names (a mere $50,000 investment) with URLs like cnn-news.net and ESPNsprts.com. When you buy a domain, you can put in anyone's name and email address as the owner, so they just filled in things like "Apple Computers," and "Time Warner" as the owners. This made the search engines and the advertising networks think they were legitimate. Then they put video ads on those domains (ads from actual companies like Google AdSense and Yahoo).

The final step was to build a massive army of virtual robots who would visit random sites as well as the fake sites. The "bots" would randomly click on parts of the page, navigate to new pages, and act like human beings, all to hide the fact that they were not people. Every time the bot clicked on an ad, the hackers would get a little payout. This is how online ads work. You get paid every time someone clicks on one of the ads on your site.

Because the ads in this case were video ads, and the domains and registered owners were "high-value domains," the payouts were much bigger than normal ads—as much as $10 to $13 per click! Even more devious, the hackers knew how the ad networks assign a value to an ad. Ad networks do this with an automated bidding system. If your ad targets "sky blue monkey training services," you're most likely the only name in the game. If you're trying to buy an ad for "lower your mortgage," you're going to pay through the nose.

The hackers built additional ads in the network to compete with the keywords they were displaying ads for, artificially raising the value of the ads they were showing. Morality aside, you can almost appreciate the criminal mastery here.

How does advertising make money?

This tale of evil genius illustrates the three key aspects of making money through online advertising.

1. The Ad Networks

If you have a website or an app that generates a fair amount of traffic, consider putting advertising up to make some cash off your hard work. You could always get a sales team together, hit the streets, and sell advertising space to anyone who will listen. This is a very difficult proposition (ask your local newspaper). As a result, ad networks run almost all Internet-delivered advertising.

The ad networks clear houses for advertisements. Say you want to advertise your local plumbing service. First, you create an account with your favorite ad network. Google AdSense is the 20,000-ton gorilla in this space, so I'll assume you went there first. Next, you upload the image (called a banner ad) or type in the text you want displayed in your ad. The system offers you all sorts of options about which users you would like to see this enticing new deal for clogged drains. This might include demographic data like age, gender, and income. Next, you provide the geographic radius, country, language, or even operating system of your target audience.

If you're thinking to yourself, "Wow, that's creepy. . . . How does Google know so much about everyone in the world that they can offer me these options?" you are not alone. Why do you think Google has such a great search tool? For the betterment of humanity? Or so it can collect truckloads of data about what you like? No judgment. Take your time.

Finally, you receive an estimate of how much you pay for a given "click"—the amount Google bills you for each time someone clicks on your ad.

Millions of website and app builders sign up on the other side of the market. The network provides them with a small piece of code that they embed onto their system. If a user who fits your targeted "male between the ages of thirty-five and fifty-five living in the greater Lancing, Michigan area who read an article on 'how to fix a sink'" visits a sporting goods site with that code embedded on it, they'll see your ad!

2. The relative price of a click based on demand

That price given to you is only an estimate. How does the network come up with that number? How do they know what your click is worth? Enter the murky world of real-time automated bidding systems.

Every time a page with an embedded ad network's code gets viewed, an amazing amount of back-room deals occur in milliseconds. First, the code sends back the user's demographic data, search terms, browsing history, physical location, and the page they're trying to view to the network's servers. Based on the preferences set by all the advertisers vying for clicks, the ad network performs an auction based on:

The maximum price an advertiser set as their upper limit

- The total daily amount spend set by the advertiser
- The quality of the match between who the advertiser's target is and the real human being loading the page
- And a bunch of other things that are super complex, secret, or plainly rely on voodoo (we'll never know because no one will tell us)

The winner gets their ad displayed to the user. If the user clicks on the ad, the winner's account gets debited, and the ad network takes their cut. The owner of the website also gets a small credit for driving their user to another site.

To put this all in perspective, Google makes around $80 billion a year from people like us clicking on little boxes on random websites.

3. The number of impressions or clicks

If you're the owner of the app or website showing the ad, how much money do you make? Well, you get paid the CPC (cost-per-click). This is a percentage of the bid rate that comes out of the real-time auction. Each time one of your users clicks an ad, you get paid.

Another model is CPI (cost-per-impression), meaning you get paid every time your ad shows up. These payouts are usually significantly lower than clicks, but they do make sense in some situations. If you plan on advertising, only use CPI ads if you think your ad is super enticing and lots of people will click it. This is because you only pay per thousand impressions rather than per click. If you think the ad is a runaway success, CPC becomes really expensive really quickly. CPI gives you a better bang for your buck.

How much money can you make?

Getting down to brass tacks, can you make money selling ads on your site or app? Well, yes, but you need a lot of traffic. On average, you will get one click of one ad from 1 percent of your users. The math is pretty easy, then. If you are getting a $1 per click payout, you'll make on average a penny for each user who visits your site. Let's put that into perspective.

There are about 300,000 sites in the world that get 100,000 unique visitors a month. There are probably only 50,000 that get a million or more a month. Keep in mind that those are unique visitors, not page views. Ads display on pages, so we want to know the number of pages displayed. A healthy page view number is two to five pages, so we'll split the difference and call it three pages per user.

We can assume that top sites can generate anywhere from 100,000 page views a month to 100 million. That's a huge gap, but we're in a very fuzzy realm, here. With a little web scouring, I found some claimed revenue numbers associated with claimed daily unique visitors.

Daily Uniques	Monthly Revenue
13,100,000	$100,000
150,000	$10,000
70,000	$20,000
1,000,000	$60,000

As you can see, because of the fuzzy nature of bid prices, industry niches, and content, there is no easy way to draw a correlation between unique views and actual revenue. The lesson here is that you need a massive amount of traffic in order to make money purely off advertising network dollars.

Private Ad Networks

As with everything in the world of business and computers, there is a caveat. If your site is fairly niche and generates a good amount of traffic, the standard ad networks are probably not what you want. In these situations, you might want to create your own sales teams or use a private label seller. These companies deal in getting advertising on your site for specific buyers. These deals are usually negotiated person to person. An example might be a sports team's fan page—it might have thousands of dedicated supporters visiting constantly, all drawn together for a specific purpose. As a result, the value of the ability to target those particular users would be better calculated in a traditional sponsorship or advertising deal rather than allowing algorithms to work out the cost-per-click. These deals are significantly more profitable than using the automated ad networks.

Selling Your Data

When all other revenue models fail, there is the old "sell my data" paradigm. The basic premise here is that your application created a large amount of data that is valuable to a third party. This can be extremely profitable, but it's a very tricky mountain to climb.

There are three main types of data that interest other entities: usage data, content data, and user data. I'll touch on these briefly, but each one has challenges and strengths.

Usage Data

Usage data is the analytics of your mobile or web application. Companies that buy this type of data look for how many users you have, their location, which social networks they log in from, and what cell carrier they use. This doesn't sound super valuable, but it helps advertisers and market researchers work out trends in a particular digital space. Usage data sells in a passive stream, meaning you integrate some sort of block of code from the purchaser, and your stats are set in a continuous stream to the buyer. You earn an incremental fee per click or tap, or per user, or per "event," whatever that might be.

Usage data sales are not a super high ROI because you need a ton of users to make it anywhere near worth your while. Usage data is also widely available and, as a result, not extremely valuable. On the positive side, selling your usage data is not very controversial. Your users won't feel a severe invasion of privacy. It is anonymous usage data with no personally identifiable information. There are a number of online services for selling usage data like Exapik (http://www.exapik.com/) or Flurry (https://developer.yahoo.com/flurry/).

Content Data

The second type of data sales out there is content data. Content data is a wide category including:

- User ratings and reviews
- Text content
- Statistical data

- Sales numbers
- Databases and lists

That's just to name a few. To give a concrete example, imagine if you had an app that lets users track their wine collection and gives them the opportunity to rate each wine they try. One way to monetize the app like that is to sell the data to a wine marketing company or a liquor store chain. Your buyers would be able to tell what types of wine people preferred in different parts of the country, or which wines they should stock due to increasing popularity trends.

This sounds great, but it's fairly difficult to do. Trying to find those partnerships takes just as much time as most enterprise sales. Your time might be better spent selling subscriptions or wine tasting kits or whatever else helps turn a profit. If your application produces huge amounts of data, there are clearing houses for specific types of content data. Infochimps (http://www.infochimps.com) is a data broker that handles large data sets, and the newly launched Data & Sons (https://dataandsons.com) allows for marketplace data sales similar to an app store.

User Data

The most controversial type of data sales is user data. By user data, we generally mean information specific to an identifiable person. This is where Acxiom (http://www.acxiom.com/) and Experian (http://www.experian.com/) play, but there are a few other smaller, equally scary data brokers out there. It's because of these guys that Target is able to send you coupons for prenatal vitamins before you even know you're pregnant (seriously, that happens).

If you have a lot of user data, especially personal purchasing data, you can make a significant amount of money with it. Ever wonder why stores love to give you discounts with loyalty cards? They more than make up for the discounts by selling your data to the big brokers. Depending on what you read, your personal data is worth somewhere between $5 and $300. Big data companies shell out a few bucks to find out that you like Pepsi over Coke, but fork over a couple hundred to find out that you suffer from chronic depression or bulimia. This is where you, as a developer, must battle with the moral decision as to whether you should protect your users or see them as commodities. You can most likely tell where I stand on this.

CHAPTER 12
MARKETING YOUR APP

Startup and tech news publications are everywhere these days. They're part of the Silicon Valley hype-machine that fuels the insane amounts of West Coast venture funding. We read TechCrunch, Product Hunt, Hacker News, WIRED, and even Forbes, and we see that CoolTech, Inc. raised another $10 million.

So, why wouldn't we want to get in on it? Regardless of what the perception in the media is, not every tech company reaches 10x growth month after month. Hitting a million users overnight is pretty rare. In fact, it's almost unheard of.

The percentage of tech startups that reach a successful exit (selling to a larger firm, going public) is minuscule compared with the number of businesses started each year. A commonly quoted business statistic is that only 10 percent of new businesses are still in operation after their second year. That number is much lower for technology companies, but why?

For some reason, there is a perception that tech businesses (or projects) are easier and quicker than other types of businesses. This might come from a lack of experience or from reading about teenagers who become millionaires overnight by writing some sort of revolutionary app. This is simply not true.

A lot of the time, the technology is the easy part. This is your manufacturing department. Instead of building hairdryers in China, you're building an app in California. Whether it's soup spoons or software, the hard parts are the same: marketing, legal, sales, customer service, supply chain, and management. Product development is the fun bit, for sure (which is why I do it for a living), but it's easy compared to dealing with people, lawyers (distinction intended), or the government. Because of this disconnect between the perceived required effort and the actual effort required, people lose heart. They get frustrated or angry, or they quit. This is not a road to success in anything.

I admit that exceptions exist. The market size and speed of software also means that on the off chance something does go viral, it can grow exponentially— which no other market can do. I can't open a food truck tomorrow, and have a million customers in a month, regardless of how good my spicy Mexican-Thai pizzas are.

The launch of a new tech product or service requires similar levels of effort as it does to launch any new brand. While your costs might be lower building an app than a brick-and-mortar store, the marketing costs remain similar or even higher. In this section, we'll go over some of the basics of marketing your new tech product or service, primarily looking at how to attract new users, and addressing a few concepts you're likely to encounter.

Internet Marketing

The Internet is a tool for social change, the destination of ideas and knowledge, and a great way to find cute pictures of cats. It revolutionized the marketing industry. There is no sector of the economy not heavily plugged online. Some of these campaigns are elegantly done, clever, beautiful, or value-adding to an insane degree. Unfortunately, most of them are just bad or silly, and some of them are downright scams.

This is such a huge topic—one with entire sections of bookstores dedicated to it. I'd recommend Digital Bacon (http://digitalbaconbook.com/) by Alex Rodríguez if you're looking for a deep dive into the subject. What I'll touch on here are some of the larger technical topics you might run into, especially when trying to attract users to your new product or service.

Search Engines and Optimization

One of the most common methods to get traction for your digital product (or any product these days) is through optimization. This is basically the idea that if you do enough of the right things, in the right order, you will come up as number one on Google when someone searches for your niche. If I have the most optimized sea monkey delivery business in the world, and you search for "brine shrimp are not monkeys," you will see my site at the top of the list.

There are a few different types of optimization, but this is a survey, not a master class.

SEO

The most common type of optimization is search engine optimization or SEO for short. SEO is a technical exercise. SEO is how you make changes to your website to make it more visible and better ranked with Google. I understand that there are other search engines out there, but Google is the only one that really counts at the moment.

The theory of SEO is something like this: Google maintains an algorithm that they use to rank your website for any given phrase (in every human language in the world). This algorithm is secret, of course, and for extra fun, Google tweaks or revamps it all the time.

Imagine a points system for your website. Google awards points for everything you do right and takes away points for everything you do wrong. Google awards for things like:

- Having the search keyword or phrase on a page multiple times
- Easily readable text (based on things like word complexity and sentence length)
- Finding the search phrase in the URL of a page

There are dozens of these rules that industry "experts" recommend. Knowing what tactics are safe can be difficult, but we'll look at a few brief tips to help you stay clear of the scams.

Things to do:

- First, set up a Google Analytics (https://www.google.com/analytics/) account. It's free and amazingly powerful. GA shows you what your visitors look at and how they come to your site.
- Check your site's "load speed" (make sure it runs quickly). Pingdom (http://tools.pingdom.com/fpt/) has a great tool to help you out with this.
- Use Google's Keyword Planner (https://adwords.google.com/KeywordPlanner). It's a great way to gauge the competitiveness of various keywords and phrases.
- Check out the competition. Find out who is the top dog in your market and run the Keyword Planner against their site.
- Write your content for a specific audience and a specific keyword. Don't combine too many ideas into one page.
- Make sure your pages' titles are explicitly about the content. I know that sounds silly, but it's amazing how many people don't do it. If your page is about gluten-free candy bars, that's what the title should be, not "Best Chocolate Bars for the Health Conscious." The Internet is a big place; people will click on the first thing that matches their search.
- Talk about your subject a lot. Content density is king, so if you're selling the most comfortable bicycle of all time, have a few pages

about comfort and bicycles. You don't want to use the phrase "most comfortable bicycle" to an unnatural extent, but you do want to use it a lot. Sprinkle in other variants as well.

- Have relevant internal links. Add links in your content to other pages on your own site. This helps search engines see the relationships between your content pieces.

- Have relevant external links. Add links to other sites you don't control. This helps the search engines see how your content relates to other sites on the web.

A lot of bad information exists about SEO techniques. You most likely receive marketing emails or have seen websites that promise things like "guaranteed front page of Google." Avoid these people like the plague. They can't help you and will likely cause irreparable harm.

Things to avoid:

- Stay away from "keyword stacking" or "keyword stuffing." This is where you use the same phrase over and over again to try to game the search engines. Google punishes you for this.

- Don't duplicate content—this used to be the mainstay of the "grey hat" SEO people. Google punishes you these days for creating mirror websites where you repost the same content and link back to your main site.

- Avoid "link farms." These are sites that exist mainly for you to be able to easily get your web address out there. It's fine to write a blog about your product on Medium.com and link back, as long as it's original content. But don't post the same thing on 100 other shady blog sites and expect to get away with it.

- Stay away from SEO companies who guarantee results. They are most likely using some of these dark techniques. You may see short-term results, but in the long run, you'll get hurt.

Search Engine Marketing

SEO is not SEM (search engine marketing). This is an important distinction. If someone tells you they're the same thing, treat that "expert" with suspicion.

What exactly is the difference?

Search engine marketing is more like traditional marketing, where you try to market your brand either through more online visibility or paid advertising. Most SEM folks out there try to get traffic to your site that will be long-term and mostly free. That sounds great, right? Well, it isn't easy and can be expensive.

Non-advertising SEM is about getting other sites out there to send you traffic. The old-fashioned way of doing this at scale was guest blogging. This is a technique where you write some content for another site, because they need content, and include a link back to your site. The problem is, Google cracked down on this. It now negatively impacts your search scores.

Savvy SEM experts turn more and more to social media to drive traffic to their sites. Microsites are another useful but expensive tool in traffic generation. You create small, entertaining websites that offer some sort of value on their own, and add links back to your main site. This is a costly and short-lived win, but it can prove worth it if done well.

If you have a product or service that you're seeking to promote, the old ways are still the best ways. Reach out to reputable sites, experts in the industry, journalists, and social media thought leaders. Try to get them to write about you. This usually means a free trial or a singing telegram if you want to get creative. When it comes to marketing something like an app, I find that it's really easy to get people to write about you with a simple technique. Find a list of all the bloggers and journalists out there who wrote about similar products or services like yours in the past, and send them a short email saying something like:

Dear Mrs. X,

I have been a big fan of your rice cooking blog for the last few years. It's something I've checked over and over again because I'm in the process of developing a rice cooker myself. Your insights proved efficacious in developing my Super-Rice-Cooker, and I wanted to say thank you for all your excellent work.

My Super-Rice-Cooker hits shelves in about a week. Can I send you one to try out? I'm not asking you to review it or give me any publicity, but I'd really love your feedback.

Please let me know if you'd be up for that, and I'll send you one of our demo models for you to keep.

Thanks so much,

Sycophantic Marketing Guy

It's a little trite, I know, but it works! People love to hear that you think highly of them. If you're asking someone for their expert opinion, they're usually eager to prove to you, the fan, that they're worthy of your praise. The thing is, they have column inches or blog pages to fill. Seeing as they're looking at your shiny new product already, nine time out of ten, they will write about it.

Ad Buying

Most of the top returned searches on any search engine are "paid placement" ads. The biggest player is, once again, Google, with their AdWords program. AdWords allows you to create ads that will run on Google's search results page, as well on any approved AdSense site. AdSense is a program where you can sign up to have Google's ads run on your own site, and they'll pay you per click or per impression.

Ad buying is a tricky business. In almost all modern ad systems, bidding algorithms determine the price of your ad. If you're buying an ad to show up when someone searches for "insurance" or "loans," you're going to pay through the nose for each click (upwards of $50 every time someone clicks on your ad). If you want to target a phrase like "banana-flavored tree frogs" you're probably not going to have any competition, and the cost-per-click will be pennies. Google's

Keyword Planner (https://adwords.google.com/KeywordPlanner) can give you insights into how to balance specificity over costs. It's easy to use and can give you a huge amount of insight into what to target. To learn more about AdWords and Google advertising in general, I highly recommend Brad Geddes's great book, Advanced Google AdWords.

Facebook is the other up-and-coming advertising giant. Facebook's ad buying program improved drastically in the last few years. Their targeting improved to the point where you can tell Facebook to display your ads to extremely specific users, for example, "women between the ages of thirty-two and thirty-five who like motorbikes and live in Daytona Beach, Florida." This might not yield a huge amount of potential clickers, but there is very little else out there that can precisely target your demographic.

My personal reason for using Facebook advertising is tracking ROIs for user signups. If I build an application that allows users to rent kittens by the hour, I can create a campaign that targets my exact demographic. Then I'd see how many clicks I get and compare that to the number of downloads and signups the app receives. I could run a few different ads, all without spending a lot of money, and work out what combination of age, gender, location, and interests will attract the most users. From there, it's an easy calculation to find out my user acquisition costs. If I know that every $5 of ad spend generates one new user, I can take that to investors and draw a straight line for them as to how much money I need to reach a critical mass of users.

App Store Optimization

App Store Optimization (ASO) is the newest specialized marketing kid on the block. ASO is the art of trying to make your app show up in searches on the Google Play Store and the Apple App Store. This has become a hugely important subject in the last three years, as the wild west of app marketing opened wide. There aren't a lot of great resources out there yet on ASO, but there are new books and sites about the subject popping up all the time, so try out your Google-fu and see what you can find.

A few rules of thumb:

- Pick a good, descriptive title. App searchers are even less patient than their web counterparts. If your app connects like-minded snake oil collectors, don't call it "Slithering Connect."

- Be distinctive. Check out the competition, and don't use a name that will confuse your app with someone else's.

- Use short titles. Keep the name of your app under 20 characters if you can. This will mean that your app's full name is never shortened by those horrible ellipses.

- Pick good keywords. App stores allow you to add keywords to your searches. Use five or less.

- Get your app rated. Ratings are important, as are reviews. Never pay someone to do this. That will get you kicked off the store in no time. I'm sure you have at least twenty people you know who you can arm-twist into giving you four to five stars and a few lines of non-hyperbolic text. Even better, write the twenty reviews yourself, send them to individuals, and ask nicely for a review.

- Get downloads. This is a little tougher, but apps with more downloads get ranked higher. Once again, don't pay anyone to do this, or you will face the wrath of Apple. Leverage everyone you know to get the app.

- Write good descriptions. Be . . . well, descriptive. Explain exactly what the app does without being excessively flowery or over the top. Try to insert keywords into your descriptions in a natural way. Bullet points are totally okay as well!

- Use interesting images. Your screenshots are critical. Keep in mind that they don't have to just be images of the app itself. Get creative with colors, people, phone frames, or pretty much anything. LaunchKit (https://library.launchkit.io/) provides a great tool for making engaging screenshots that will capture your searcher's eye.

Social Media Marketing

Social media marketing is the process of leveraging social networking sites to get organic traffic to your web or mobile application. This is a content media marketing technique similar to the SEM section above; post high-quality and engaging content on your social media pages and hope it generates traffic. Unfortunately, this is not as easy as it sounds.

The key here is engagement. You can't post anything that comes to mind and then hope for the best. You need to engage with your users. Respond to tweets, comment on Facebook posts, and like things on Instagram; pretend you're a teenager! Corey Perlman's Social Media Overload: Simple Social Media Strategies for Overwhelmed and Time Deprived Businesses is a great place to start.

When starting your social media marketing campaign:

- **Have a strategy.** You need to set up a specific, targeted strategy. Find out where your potential users hang out. Don't rely on LinkedIn engagement to market your online movie trivia game!

- **Create value.** Don't always occupy advertising mode. Answer questions on Quora, comment on other people's posts, tweet back at people asking good questions, and share other people's posts.

- **Share, like, and comment.** Do it often, at least a few times a day. Keep in mind to stay focused on the message created in your strategy. Don't create noise.

- **Attract followers.** Adding good content and staying engaged will help you grow your follower base. Good old-fashioned networking is also extremely powerful. This is especially true if you can attract a few key influencers—people who have a lot of followers themselves.

- **Follow the right people.** Find influencers who can drive traffic for you, and follow them. Engage on their sites (while avoiding coming off as a stalker), and you'll find their followers start getting curious about you.

Traditional Methods

"I have always believed that writing advertisements is the second most profitable form of writing. The first, of course, is ransom notes. . . ." —Philip Dusenberry

Search engine optimization is not the end-all, be-all of marketing your software. There are libraries out there on marketing techniques, advertising strategies, and sales models. Don't disregard traditional marketing methods when it comes to non-traditional products. There are caveats, however. I once worked with a team who spent 20 percent of their initial investment on the software development and the remaining 80 percent on billboard ads in Northern Michigan. The ad was the name of the website and an attractive lady in a bikini. That's it. . . . Just the lady, her swimwear, and the URL. The site was a directory for finding contractors. I'll let that sink in. They didn't do very well.

Scantily clad billboard blunders aside, there are some marketing concerns that are more central to software products and services than traditional businesses. Search and advertising are still king, but word of mouth is extremely valuable in certain markets. Reaching out to potential users beyond the digital world can drive real traffic to your new software product.

Traditional Methods and Launching

Launching any product is tough, but certain business models are significantly trickier. The most difficult model is the platform (or marketplace) concept.

The upside of these systems is huge because they can become self-sustaining and highly profitable for the implementer or middleman in charge of the platform itself. However, any system that requires buyers and sellers—or content creators and content consumers—or any similar scenario, will double your marketing headache because you now have two distinct groups you need to woo.

There are three main tactics for building up this kind of business:

1. **Target one side first**. If your platform can somehow survive with only one side of the ultimate goal, you cut your workload in half. A good example of this is LinkedIn. They created a professional social network that attracted their non-paying user base of people who wanted a personal business profile. Once they had enough users to make the

numbers work, they opened up the revenue side of the business to recruiters and job posters.

2. **Get a few big clients first**. Become the go-to platform for a company that people already want. Your first clients will drive users to your service, and that will attract other customers once they see the big boys using it. A good example is Glassbreakers, an employee engagement and talent development tool. They built their application specifically to solve the needs of a few large tech companies like Box and Pinterest. After word got out, everyone and their millennial workforce wanted in.

3. **Go for both sides at once**. This is the trickiest strategy but can ultimately be the most successful. Here, you need to attract both sides at once or try to pick up one supply side and one purchaser side at the same time. Freelancing service Upwork started this way, as did our next big player: Uber.

Uber needed people to sign up to be drivers when there weren't any passengers, and they needed users to sign up for rides when there weren't any cars available to pick them up. This is your standard "chicken or the egg" problem. So how did Uber get over this hurdle? They went for both sides at the same time and primed the pump with cash. They did this by paying drivers before they had passengers and then gave away free rides on top of that. It was an expensive game, but it paid off.

This "cheating" by paying both sides is fairly common in the startup world. PayPal had an early user acquisition cost of roughly one million dollars a month! This isn't the kind of money you're likely working with, but it illustrates a point. PayPal paid $20 to each of their new users and another $20 if they referred anyone. A user acquisition cost of $40 direct is hard to wrap your head around, especially considering they grew at around 100k users a month. Elon Musk contends that PayPal's initial spend on user acquisition was between $60 and $70 million.

If you've got a multibillion-dollar idea and can convince someone to give you $70 million to prove it out, you most likely don't need this book (but call me). The idea of paying your future customers to use your product or service is a potentially winning strategy.

Buying users doesn't have to come in the form of giving them cash. Getting teams of people to manually sign users up is another pay-for-user strategy. This means real people (or interns, if interns don't count as real people) hitting the streets, work sites, clubs, or wherever your target audience is, and getting them to sign up on an iPad. It's grueling work, but it pays off. Gaining users, regardless of how you get them, proves "traction" to stakeholders and investors. "Street Teams" or "Ambassadors" are extremely successful in social-based startups and are worth a look.

Traditional Media Buying

With unlimited budgets, you can go with real-world advertisements on TV, radio, subway cars, and yes, even billboards. These methods are mainly useful for cementing existing brands or creating attention for a new product line. If you saw a TV spot for a new music streaming service from Facebook, you might check it out the next time you were online. But it's highly unlikely that you would remember a web address for a new music service you'd never heard of if you saw it on the side of a bus.

There is an argument to make for media buys when they relate closely to your target market. If you're launching a cigar reviewing app, it might make sense to buy a page in *Cigar Aficionado*. This is a fairly common sense decision in most situations.

CHAPTER 13
FINAL LESSONS

The failure rate for internal software projects is around 50 percent. That means that for every project you try to implement inside your organization, another will fail. That's nuts, right? It doesn't have to be that way. There are a few things to watch out for in every project—warning signs that things are off the rails. I've tried to make this list as exhaustive as possible to help you avoid these pitfalls.

Danger Signs

Developers

Software people are a slightly different breed than most. They are cliquey, insular, and often times, smug. It takes a certain type of manager to make sure everyone is in lockstep.

Trust them or toss them

When hiring a new developer or development company, do your vetting. But after you've hired them, don't get too comfortable. Start by giving them a small task to complete first. Never—I repeat, never—hire anyone to build your three-million-dollar application without giving them a $40,000 project first and judging the results.

If a developer lies to you or misses more than two deadlines early on, get rid of them. There are millions of wannabe programmers out there who are unprofessional, have inflated self-images, and have little actual expertise to back it up. Never settle for a "video of it working"—insist on access to play with whatever demo they have. Videos can be faked. Many programmers won't want to show you something incomplete, but if you insist and make it clear you don't expect perfection, you'll find an honest programmer quickly becomes your friend.

Rewriting code

There are two cases you will encounter when developing software that require a judgment call around "rewriting code."

The first situation is extremely common when moving to a new development team, hiring a new programmer, or replacing an outsourced resource. 50 percent of the time, I guarantee the new team will tell you, "This is awful! We need to start from scratch." This is the last thing you want to hear. How much is this damned thing going to cost in the end, anyway? It seems to work fine now, so why do I need to rebuild it all? Here there are two possibilities:

1. Your previous developer was actually rubbish at what they did. Your new team tells you that if this piece of junk goes live, it will end up crashing, will be hacked, or will randomly delete your customers' data as soon as there are more than five people on the site.

2. Your new developer just doesn't like your previous team's style, their programming language of choice, or the framework they used.

These two outcomes are equally likely in my experience. The only way to know for sure is to hire a professional firm to do a code review or audit. There are many reputable development companies out there that will do this. A third-party review will give you an honest report as to whether or not you need a rewrite or if you need a more open-minded developer.

Internal and External IT People

Early in this book, I warned you about the dangers of lumping all tech people under the generic label of IT. I bring this up again because it is extremely common to place the development of a software project under the IT department's supervision. IT departments are extremely important, and I doubt I could get a job in traditional IT. Software developers rely heavily on IT infrastructure in the same way toilet designers rely heavily on plumbers. My standard joke is, "At the end of the day both sides are dealing with crap but from an entirely different perspective."

Rather than heading straight for the IT department, I recommend starting software projects out of your marketing department (if you don't have a dedicated engineering team). If you do have an in-house software development department, make sure they are not under the IT department's supervision. Culturally, and technologically, that would be like having your creative department managed by accounting.

Scope Creep

Scope creep is the process by which a project becomes bigger and bigger because of outside stakeholders making additional requests before the application can stand on its own.

Scope creep is the biggest killer of morale and of projects in general. Developers, like everyone else, prefer to work in manageable chunks of work that provide discreet victories. There is nothing worse than completing two weeks' worth of work, ending in a demo of your new cool feature, and hearing, "That's great, but can we make it do X?"

This scenario is common when you have a weak project manager who can't say "no" to the project's stakeholders. Try to break the project down into the smallest possible component parts. Start with the smallest feature. If this is your first foray into development, it might be better to eat the elephant one bite at a time, rather than trying to do it all in one sitting. Once you have the first tiny bit done, start planning the next piece. This type of phased development is not only useful for proving concepts, seeing progress, and celebrating small victories, but it will also massively improve your relationship with your development team.

Complicated vs. Complex

Most software problems come in three flavors: Complicated, complex, or both. While the difference may seem subtle, there are important distinctions.

Complicated systems imply lots of moving parts. You can describe most software projects this way. These parts are knowable or learnable. We can sit down with all the experts and stakeholders, map out the system, and understand how the whole thing fits together. Your accounting system might be very complicated, but with enough time and energy, you can break it down into its individual relationships. It can then be modeled and programmed.

Complexity is the measure of how knowable a system is and how predictable the outcomes are. This is where most of the cutting-edge technologies play—in the highly complex world. Your hotel booking software may be complicated, but the market drivers for demand may be complex.

Building a system that tries to predict your demand and then advertises prices dynamically based on that demand is complex.

Most people don't seem aware of these two variations of problems. Building complicated software is what most programmers do for a living. Complex systems usually require a different type of engineer or a team who can handle both types of thinking. My company frequently hires mathematicians, physicists, economists, and statisticians to help us model complex interactions. We try our best to learn how our systems work, but sometimes the math is beyond our comfort zone, so we have to trust the expert's algorithms.

This distinction is important when taking into account budgetary requirements, but it also helps understand the scope of the problem you're trying to solve.

Buy-In

The most disappointing aspect of any software project is when no one uses it. If the product-market fit was wrong, or if your users preferred the competition, it's either bad luck, an opportunity to iterate, or a lesson well learned.

If the application you built was intended for use internally, as part of your ongoing business processes, you have no one to blame but yourself if it flops. I've seen millions of dollars spent on building a company-wide customer relationship system, only to witness every salesperson sticking with their smartphones and spreadsheets. How is that even possible?

This is a potentially disastrous scenario. If this has happened to you, it might feel impossible to fix, but there is hope. Problems regarding integration can always be traced back to a simple failure to create buy-in with your internal customers.

In this example, the company in question almost never asked the opinion of the sales staff while the tool underwent development. Post-launch training was nonexistent, and the CEO never made it mandatory to use the application. Buy-in comes from the top down but requires everyone to get on board. This is really not that difficult to pull off, but it takes dedication.

Over the years, I've come up with some key points in helping to get this done. Rather than blabber on about the problems and costs, I'll share some of these points with you now.

These are things that you as a project manager—or, even better, a senior manager—should think about and probably do!

1. Prioritize an effective buy-in strategy while making technology and software choices.

- Work with shareholders in the company when choosing your technology to ensure that users, needs, and tech align. Do Baby Boomer managers need or want to carry around an iPad? Or does a Millennial IT manager just think they do?

- Guide the organization on complexity design decisions before purchase.

- Do they want or need minimal buy-in software (i.e., easy to learn, easy to use, and easy to re-learn next month) for tasks that are only done occasionally?

- Assess software or tech options for ease of adoption and recommend the most suitable choice on the basis of company culture and experience.

- Assess other factors in user adoption: training materials, manuals, and help files for user-friendliness. Review issues such as the software/ tech provider's training and after-training service, back-up history, accessibility, and reputation for service.

- Assess the real cost of training to adoption.

2. Guide the pilot project. Ensure the involvement of the right people, and obtain the right data.

- Create a list of all stakeholders.

- Schedule a project check-in cadence.

- Perform milestone reviews with everyone involved.

- Collect feedback in a central location to which everyone involved can refer.

- Create a feedback loop, accept comments, and implement a percentage of them to prove that you're listening.

3. Effectively define the training objectives and user goals.

- Communicate training objectives and goals clearly and in a timely manner to users.

- Determine what the users know before training versus what they need to know (i.e., how big the gap is in total skill sets).

- Determine the minimum degree of functionality each user group should reach, sustainably, to achieve the desired business objectives.

- Determine whether it is possible to create levels of functionality so that certain users can achieve adequate functionality, while training for other functionalities later, according to confidence, need, and interest.

- Assess and monitor the point. A blanket demand for full functionality can lead to a significant percentage of any user group feeling "overwhelmed with feeling overwhelmed."

- Create ways to show options to learners so that they can choose their level of functionality before or during training.

Determine these tasks during the pilot project phase.

4. Identify sources of concern.

- Roadblocks

- Resistance, "out of comfort zone"

- Anxiety

- Problems with degree of management buy-in to the system

- Degree of senior management support

- Concerns with training methods

Work through any issues. Any engagement requires (amongst other things):

- Involving people

- Listening to them

- Demonstrating that their opinions count

- Making sure that people feel successful

- Communicating—people always know what they are doing, why, and their expectations

- Giving them feedback

- Communicating in terms that people understand and at a conceptual level matching their phase of understanding.

5. Assess or monitor training to ensure that it reaches a tipping point where the goal percentages of users reach and maintain the defined adoption standard.

- Ensure that training accords with the laws of adult learning.

- Ensure that training accords with the users' needs and styles of learning.

- Ensure that training achieves specific goals, along clearly mapped learning paths.

- Ensure that training includes teaching the users of the software/technology to think of beneficial ways to use the new tools in their own daily work.

- There are experts in this. Don't think that because you are such a great programmer or project manager, training people or designing retainable classes will be easy.

6. Maintain use of the software/technology.

- Create a supportive framework for the implementation phase—with top-level, go-to mentors (who can also act as system champions) and competent buddies (pairing buddies with slow adopters and resisters). Consider involving the top-level mentors in the training itself.

- Ensure that IT staff are supportive, that IT staff works with users at their level using vocabulary they understand, and that they buy in to the mission of engaging the users with the new material or equipment. Provide the IT department staff with incentives and reward them for meeting the adoption and sustainability goals of the project.

- Work with management to tie the technology to the key performance indicators of each position in a way that the use of the technology is mandatory. Ensure that all managers use the technology or receive coaching if they themselves are slow adopters. Help managers to (a) reinforce the "what's in it for me" factor for the users, (b) reward and give feedback on adoption, and (c) apply a consistent, no-exception policy to use the software/equipment.

- Schedule ongoing training until reaching an agreed-upon level of adoption. Ensure that the company invests to the degree that the technology is critical to the business, in regular refresher bursts, assessments, rewards, and games.

- Celebrate early adopters and ensure praise of others as they meet their goals.

7. Measure the results of the implementation and training against the business goals.

- Assess the pilot project and identify/report problems and solutions.

- Survey users for frustration, comfort, opinions, problems, and successes during implementation.

- Measure the results of the project in terms of the use of technology/software and the achievement of business goals you put it in place to attain. (With project management software, one might measure the number of overruns or schedule misses since implementing the new program).

I know that is a lot to go through once you've already gotten so far into a project. However, if you carefully manage each aspect of the steps of creating buy-in, your project will have a much higher chance of succeeding.

Each one of the pitfalls in this section can sink an entire project. The biggest one to watch out for is making sure you have a solid development team whom you trust. After that, the hard part is managing your internal team to make sure your IT staff doesn't get too territorial, and your internal customers have enough buy-in that they will be excited to test the new product out for you.

Conclusion

As we come to the end of this book, my hope is that you would now feel comfortable walking into a programmer's office, or through a development firm's door, with a good understanding of the technologies, processes, and terms that will be discussed. You should now be able to tell how to choose a strong development team and hire good programmers. You should also have a much better idea of when someone is trying to pull the wool over your eyes.

So as a conclusion to this book, I have summarized the points that you need to take away from Herding Cats and Coders in order to build a successful application.

Purpose

Make sure you're doing this for the right reasons. Building software is not something that should be taken lightly. If you start out thinking, "I'll just have to spend a few thousand dollars to get something built that will take off," you're setting yourself up to learn a painful lesson.

Participation

Programming is part art, part science. Software development is also a collaborative sport. You need to participate, and you need to be involved a lot. You don't have to understand every aspect of it in order to be on the same page as a developer, nor to call bullshit. What you must do is arm yourself with some knowledge of the practice in order to make sure you're playing by the same rules, so you can participate more. It will be fun if you know what is going on, or at least will reduce a lot of the frustration. Your developers will respect you more, and you will be able to hold them accountable.

Planning

Don't skimp on planning. All developers being equal, engineering and understanding the problem is more important than the code. Keep in mind that no plan survives contact with reality. So even if you've done all the planning in the world, things will need to change. Stay flexible, and test your assumptions as often as you can. You can always refer back to the contents of *Herding Cats and Coders* throughout the development process to see what should be happening next.

Remember, you should pay for planning—up to 30 percent at least of your total budget. Get everything designed on paper first, validate it with your stakeholders, and use it as your source of truth in development contracts.

Design

Make sure you think long and hard about the UI and UX. You should probably consider hiring a secondary party here. This will give you another person to help manage your development team and bring an outside party to play "guide."

Design is more important than you think to the success of any system. Your programmers will most likely think that the design isn't as important as the functionality, because that is how most developers think. Make sure you are design-driven throughout the process.

Technology

Choose a technology that is right for your budget and organization. Consider open-source technologies because they're cheap, high-quality, and make the world a better place.

You can build applications for the web and for mobile all on the same platform, using basically the same technology. Don't be wasteful, and abstract as much as you can.

Budget

When you hire developers, you will get what you pay for. If you are going to build something to run a $10 million a year company, you will not be able to get away with a $20,000 custom application. Rather, you should look at your development budget as a monthly spend and look it more like rent or payroll. Spending 10 percent of revenue on R&D or continual improvement of software is not outlandish. Talented engineers are not cheap, so if you're shopping around and find a really good deal, be warned.

Sales and Marketing

Engineering and design are not the hardest parts of building a company or product. The most difficult parts of business are always sales and marketing, relationships, cash flow, and customer acquisition. Building the actual product is tricky, but it should never be the most difficult task. Focus on the things you can control: planning, customer acquisition, and creating buy-in.

If you are going to market your application to the general public, make sure that you have at least triple your budget set aside for marketing efforts. Gain traction anyway you can, even if it means going down to a university and offering students ten bucks each to sign up right there on your iPad.

Fun

Finally, building an application is ultimately an act of creation. It is more like producing a small movie, where you need to be responsible for carrying the vision, but also for hiring the writers, actors, a director, camera operators, caterers, and visual effects staff.

At the end of the day, there should be some level of joy in the process of creation. Seeing something come to life and be transformative in some way is a pleasure that I hope you are able to experience. It is that feeling, that joy in seeing new projects come to life, that keeps me doing this job. I spend about half my day looking at new systems coming online and saying, "Wow, that is so cool!"

To be fair, the other half of my day is usually filled with "That wasn't supposed to happen," or "Hmm . . . I think I found a bug." But, hey, it's a living!

ABOUT THE AUTHOR

Greg Ross-Munro is the founder, CEO, and senior partner of Sourcetoad, an enterprise software development consultancy. Greg is an expert in large-scale system design for enterprise organizations. He also works on human systems interaction, business process design and optimization, the software development life-cycle, and software startups and business models.

Greg earned a BS in industrial psychology, an MBA in information systems management, as well an MS in entrepreneurship in applied technologies from the University of South Florida.

Greg has been programing since he was 8 years old and sold his first software consultancy at 24. Since then has helped over 300 companies take digital products to market. Greg received the USF MSE Outstanding Alumni Award, is a judge for Startup Weekend Tampa Bay, sits on the board of directors of StartupBus, and has been named as one of Tampa Bay's 40 Under 40.

He delights in good Scotch whisky and suffers from a dark sense of humor, and an incomprehensible patience with beautiful but temperamental British sports cars.